SPRINGER TRACTS IN MODERN PHYSICS

Ergebnisse
der exakten Natur-
wissenschaften

Volume **70**

Editor: G. Höhler

Associate Editor: E. A. Niekisch

Springer-Verlag Berlin Heidelberg GmbH 1974

Manuscripts for publication should be addressed to:

G. Höhler, Institut für Theoretische Kernphysik der Universität, 75 Karlsruhe 1, Postfach 6380

Proofs and all correspondence concerning papers in the process of publication should be addressed to:

E. A. Niekisch, Institut für Grenzflächenforschung und Vakuumphysik der Kernforschungsanlage Jülich, 517 Jülich, Postfach 365

ISBN 978-3-662-15542-4   ISBN 978-3-540-37918-8 (eBook)
DOI 10.1007/978-3-540-37918-8

Quantum Statistical Theories of Spontaneous Emission and their Relation to Other Approaches

G. S. AGARWAL

Contents

1. Introduction

The purpose of this article is to review spontaneous emission from several different viewpoints, although a large part of it will be devoted to the quantum statistical theories of spontaneous emission which have been developed recently, and to discussing the interrelations among different approaches. Einstein, at the turn of the century, presented statistical arguments to determine the rate at which spontaneous emission occurs. A dynamical theory of spontaneous emission was first given by Weisskopf and Wigner [1] and very many questions were discussed concerning line shapes [1, 2]. The Weisskopf-Wigner theory was later reformulated in very general terms by Heitler and Ma [3, 4] and by Goldberger and Watson [5]. All these formulations have been extremely successful in atomic physics. The shifts and widths of the levels appear in a very natural way in the formulations of Heitler and Ma and Goldberger and Watson. We must add to these formulations the one by Low [6], although this is not used in atomic physics because of its complexity. Recently the Bethe-Salpeter equation has also been used to treat spontaneous emission [7].

The present interest in spontaneous emission is largely due to a classic paper by Dicke [8]; he used the usual perturbation theory to calculate the radiation from a collection of identical two-level atoms initially prepared in certain specific states. He found that under certain circumstances the radiation rate from a small sample of such atoms is proportional to the square of the number of atoms (coherent radiation rate); against this, the incoherent rate, i.e. the case of atoms emitting independently, is proportional to the number of atoms.

Interest in spontaneous emission has also been stimulated to a great extent by a series of papers by Jaynes and coworkers [9–12]. In view of the fact that quantum electrodynamics is plagued by divergence difficulties, they developed a form of the semiclassical theory of radiation in which the electromagnetic field is not quantized and the source of the electromagnetic field is taken to be the distribution of currents in the atomic system. Such a theory, although it has no divergences, predicts results which are very different from the results predicted by quantum electrodynamics. It is only recently that careful experiments [13] have been carried out to test the predictions of each of the theories.

In the usual theories of spontaneous emission questions of dynamical (time-dependent) atomic coherence and field coherence are not normally considered[1]; interest has focused mainly on the decay of the

---

1 For the case of a single two-level atom emitting spontaneously, the lowest-order field correlation function has been computed by Haken et al. [14] using the Weisskopf-Wigner method.

prepared state and the line shape. Moreover, the usual theories become exceedingly involved as the number of atoms taking part in spontaneous emission increases, partly because one has to keep track of all the relevant states involved in the process of emission. Quantum statistical theories, which we discuss at length, are especially well suited to such cases and naturally predict results identical to those predicted by the usual theories. Using the quantum statistical theories, one can easily see how to make a transition to the semiclassical theory of Jaynes and coworkers. Quantum statistical theories are interesting in their own right as they are based on the techniques of nonequilibrium statistical mechanics [2], developed mainly over the last few decades.

In Chapter 2 we discuss the basic Hamiltonian characterizing the interaction between radiation and matter, and some questions concerning contact interactions and Dicke states. In Chapter 3–5 we present brief treatments of the Weisskopf-Wigner, Heitler-Ma, and Goldberger-Watson methods along with the application of each of these methods to some simple but illustrative systems. In the remaining sections (except Chapter 16) we review the quantum statistical theories of spontaneous emission and present a number of new results. The organization of these sections is as follows: In Chapter 6 we consider a collection of identical N two-level atoms and obtain the master equation describing spontaneous emission from such a system. This section also contains a detailed discussion of the interpretation of the various parameters appearing in the master equation as well as some comments on the two forms of the interaction Hamiltonian discussed in Chapter 2. We show in Chapter 7 how the solution of the master equation can be used to calculate the statistical properties of the field. Expressions for the line shape and radiation rates are also given. In Chapter 8 we discuss both the quantum and the c-number Langevin equations. The drift term in the quantum Langevin equation can be written in a form which bears some resemblance to the term occurring in Bloch equations, except that the source term in the present case is not a c-number but an operator. We first use the master equation in Chapter 9 to obtain results concerning cooperative frequency shifts and transition rates, which one could also have obtained by the use of Fermi's Golden Rule. Dicke's superradiance [8] both for small and large samples is discussed within this framework. As an illustrative application of the master equation and the Langevin equations, we consider in Chapters 10 and 11 the cases of a single two-level atom and of two two-level atoms. The results are compared with those obtained by conventional approaches (Chapter

---

2 It is perhaps of some interest to note that statistical mechanics borrowed many of the methods of quantum field theory whereas here we borrow the methods of statistical mechanics to discuss a problem of quantum field theory.

3–5). The dynamical aspects of spontaneous emission from a collection of harmonic oscillators are discussed in some detail in Chapter 12. The next two sections are devoted to the study of spontaneous emission from a small sample of two-level atoms. The resulting master equation is exactly solved and the radiation rates, line shapes, atomic correlation functions and field coherence properties are calculated. The approximate results for the radiation rate are also given. In Chapter 14 the exact solution of the c-number Langevin equations is given and the connection with a phenomenological model of time zitter used occasionally in the theory of lasers and superradiance is also established. Next, the spontaneous emission from a multilevel atom is studied by means of the master equation. The results are compared with those of Chapter 3–5. Both degenerate and non-degenerate cases are considered. Chapter 16 is devoted to the neoclassical theory of Jaynes and coworkers. The relation of this theory to the quantum statistical theory is discussed in detail and the transition from the quantum electrodynamic to the neoclassical equations is outlined. The next two sections are concerned with the spontaneous emission in presence of external fields, which may be c-number or quantized fields. Several special cases are considered. The article concludes with three appendices. In Appendix A the role of the rotating wave approximation in spontaneous emission is discussed with special reference to the question of frequency shifts. Appendix B explains the application of Mori's method to obtain the Langevin equations. Both the linear (harmonic oscillator) and nonlinear (two-level atoms) models are discussed. In Appendix C we discuss the steady-state solution of the master equations, describing spontaneous emission from the viewpoint of microreversibility.

2. Interaction Hamiltonian

It has been shown by Power and Zienau [15] that the interaction Hamiltonian between a system of bound charges and the radiation field in the *dipole approximation* can be written in the form

$$H' = H_0 - \sum_j \boldsymbol{d}_j \cdot \boldsymbol{E}_j + 2\pi \int |\boldsymbol{P}|^2 \, \mathrm{d}^3 r = H + H_{\text{self}}, \qquad (2.1)$$

where

$$H = H_0 - \sum_j \boldsymbol{d}_j \cdot \boldsymbol{E}_j, \qquad H_{\text{self}} = 2\pi \int |\boldsymbol{P}|^2 \, \mathrm{d}^3 r,$$

$$P(r) = \sum_j \boldsymbol{d}_j \, \delta(r - r_j), \qquad\qquad\qquad\qquad\qquad (2.2)$$

and where H_0 is the unperturbed Hamiltonian of the atoms and the radiation field, E_j is the transverse part of the electromagnetic field at the point r_j where the jth charge (atom) is located, d_j being the dipole moment operator for the jth atom. The last term on the right-hand side of (2.1) contains self energies and contact interactions only and does not contribute to the interaction with the electromagnetic field. It is, however, important for Lamb shift considerations. The mode expansion for the quantized electric field is given by

$$E(r) = i\Sigma (2\pi ck/L^3)^{1/2} a_{ks}\varepsilon_{ks} e^{ik\cdot r} + \text{H.C.} \quad (\hbar = 1), \tag{2.3}$$

where L^3 is the volume in which the field is quantized and will eventually be taken to infinity. ε_{ks} is the polarization vector possessing the property

$$\sum_s \varepsilon_{ks}^{(i)}\varepsilon_{ks}^{(j)} = \delta_{ij} - \hat{k}_i\hat{k}_j, \tag{2.4}$$

where $\hat{k}$ is the unit vector in the direction k. a_{ks} and a_{ks}^+ are the annihilation and the creation operators satisfying the commutation relations

$$[a_{ks}, a_{k's'}^+] = \delta_{kk'}\delta_{ss'}, \quad [a_{ks}, a_{k's'}] = [a_{ks}^+, a_{k's'}^+] = 0. \tag{2.5}$$

The corresponding expression for the magnetic field is given by

$$B(r) = i\Sigma (2\pi c/L^3 k)^{1/2} (k \times \varepsilon_{ks}) a_{ks} e^{ik\cdot r} + \text{H.C.}. \tag{2.6}$$

We first consider the interaction Hamiltonian for a collection of identical *two-level atoms*. Let us denote by $|1\rangle_j$ and $|2\rangle_j$ the excited and the ground states of the jth atom [3]. It is clear that the dipole moment operator possesses only off-diagonal elements and hence can be written in the form (cf. [16])

$$d_j = d_{12}^{(j)} |1\rangle_j {}_j\langle 2| + \text{H.C.} \equiv d(S_j^+ + S_j^-), \tag{2.7}$$

where we have assumed that with proper choice of phases the dipole matrix element can be made real. Since we assume that all the atoms are identical and similarly oriented, $d_{12}^{(j)}$ is independent of the index j. In (2.7) $S_j^\pm$ are the operators defined by

$$S_j^+ = |1\rangle_j {}_j\langle 2|, \quad S_j^- = |2\rangle_j {}_j\langle 1|. \tag{2.8}$$

We also introduce the operator S_j^z defined by

$$S_j^z = \tfrac{1}{2} \{|1\rangle_j {}_j\langle 1| - |2\rangle_j {}_j\langle 2|\}, \tag{2.9}$$

and we have, of course,

$$|1\rangle_j {}_j\langle 1| + |2\rangle_j {}_j\langle 2| = 1. \tag{2.10}$$

---

3 Throughout this article, except in the case of a harmonic oscillator, the state $|1\rangle$ will represent the uppermost state of the atom, and the states $|2\rangle, |3\rangle,...$ will represent successively lower states.

It can be shown that the operators $S_j^{\pm}, S_j^z$ defined by (2.8) and (2.9) satisfy the angular momentum commutation relations corresponding to spin $\frac{1}{2}$ value, i.e.

$$[S_i^+, S_j^-] = 2\delta_{ij}S_i^z, \quad [S_i^z, S_j^+] = \delta_{ij}S_i^+, \quad [S_i^z, S_j^-] = -\delta_{ij}S_i^-,$$
$$S_i^+ S_i^+ = S_i^- S_i^- = 0, \quad S_i^z S_i^z = \tfrac{1}{4}, \quad S_i^+ S_i^- + S_i^- S_i^+ = 1, \quad (2.11)$$
$$S_i^+ S_i^z = -S_i^z S_i^+ = -\tfrac{1}{2}S_i^+, \quad S_i^- S_i^z = -S_i^z S_i^- = \tfrac{1}{2}S_i^-.$$

With the use of (2.3) and (2.7) the interaction Hamiltonian for a collection of identical two-level atoms and the radiation field becomes in the *second quantized* notation

$$H = \omega \sum_i S_i^z + \sum_{ks} \omega_{ks} a_{ks}^+ a_{ks} + \sum_{iks} \{g_{iks} a_{ks}(S_i^+ + S_i^-) + \text{H.C.}\}, \quad (2.12)$$

where $\omega = E_1 - E_2$, the energy separation between two atomic levels, $\omega_{ks} = kc$, and the coupling coefficient is given by

$$g_{iks} = -i(2\pi ck/L^3)^{1/2} (d \cdot \varepsilon_{ks}) e^{ik \cdot r_i}, \quad (2.13)$$

which we will, at times, also write as

$$g_{iks} = g_{ks} e^{ik \cdot r_i}. \quad (2.14)$$

For the sake of completeness we also present the form of interaction Hamiltonian if the $-A \cdot p$ interaction is used (with dipole–dipole interaction included and term A^2 ignored)

$$H' = \omega \sum_i S_i^z + \sum_{ks} \omega_{ks} a_{ks}^+ a_{ks} + \sum_{i \neq j} V_{ij} S_i^+ S_j^-$$
$$+ \sum_{iks} \{g_{iks}(\omega/kc) a_{ks}(S_i^+ - S_i^-) + \text{H.C.}\}, \quad (2.15)$$

where V_{ij} is the dipole–dipole interaction

$$V_{ij} = \left\{ d \cdot d - \frac{3(d \cdot r_{ij})(d \cdot r_{ij})}{r_{ij}^2} \right\} r_{ij}^{-3}, \quad (2.16)$$

$$r_{ij} = r_i - r_j.$$

We next examine the self-energy term

$$H_{\text{self}} = 2\pi \int |P^{\perp}|^2 \, d^3 r + 2\pi \int |P^{\parallel}|^2 \, d^3 r, \quad (2.17)$$

where the transverse and the longitudinal parts are given by

$$P^{\perp} = (2\pi)^{-3} \sum_i \int d^3 k \{\hat{k} \times (d \times \hat{k})\} e^{-ik \cdot (r_i - r)} (S_i^+ + S_i^-), \quad (2.18)$$

$$P^{\parallel} = (2\pi)^{-3} \sum_i \int d^3 k \{\hat{k}(d \cdot \hat{k})\} e^{-ik \cdot (r_i - r)} (S_i^+ + S_i^-). \quad (2.19)$$

On substituting (2.18) and (2.19) in (2.17) we obtain

$$H_{\text{self}} = H_d + \sum_{i \neq j} \mathscr{V}_{ij} S_i^+ S_j^- + \sum_{i \neq j} V_{ij} S_i^+ S_j^- . \tag{2.20}$$

The last two terms on the right-hand side are due to $|P^\perp|^2$ and $|P^\parallel|^2$, respectively, and H_d is a number (not an operator) which diverges as $\int k^2 \, dk$. The parameter $\mathscr{V}_{ij}$ is given by

$$\mathscr{V}_{ij} = (|d|^2/2\pi^2) \int k^2 \, dk \int d\Omega' \sin^2 \theta' \, e^{ik \cdot r_{ij}} , \tag{2.21}$$

where θ' is the angle between d and k. On using the expansion

$$e^{ik \cdot R} = 4\pi \sum i^l j_l(kR) \, Y_{lm}^*(\theta, \varphi) \, Y_{lm}(\theta', \varphi') , \tag{2.22}$$

where θ', φ' are the angular coordinates associated with k, the z axis being along d; and the orthogonality of spherical harmonics (2.21) reduces to

$$\mathscr{V}_{ij} = 2\pi^{-1} |d|^2 \int k^2 \, dk \{ \tfrac{2}{3} j_0(kr_{ij}) + (\cos^2 \theta - \tfrac{1}{3}) j_2(kr_{ij}) \} . \tag{2.23}$$

Note that $\mathscr{V}_{ij}$ as given by (2.23) also diverges in the limit $r_{ij} \to 0$. In deriving (2.20) we have also ignored certain terms like $S_i^+ S_j^+$ etc.

It should be noticed that, if one makes the rotating-wave approximation (RWA), i.e. ignores antiresonant terms like $a_{ks}^+ S_i^+$ which correspond to simultaneous creation of a photon and atomic excitation (virtual transition), then (2.12) reduces to

$$H = \omega \sum_i S_i^z + \sum_{ks} \omega_{ks} a_{ks}^+ a_{ks} + \sum_{iks} \{ g_{iks} a_{ks} S_i^+ + \text{H.C.} \} . \tag{2.24}$$

We will discuss the full consequences of the rotating-wave approximation in Appendix A.

Finally, the Hamiltonian describing the spontaneous emission from a system of harmonic oscillators is given by

$$H = \omega \sum_i S_i^z + \sum_{ks} \omega_{ks} a_k^+{}_s a_{ks} + \sum_{iks} \{ g_{iks} a_{ks} (a_i^+ + a_i) + \text{H.C.} \} , \tag{2.25}$$

where a_i and a_i^+ are the boson annihilation and creation operators satisfying the commutation rules

$$[a_i, a_j^+] = \delta_{ij}, \quad [a_i, a_j] = [a_i^+, a_j^+] = 0 . \tag{2.26}$$

We now consider in somewhat more detail the systems confined to a region whose linear dimensions are small compared to a wavelength (ω/c). In such cases the spatial variation of g_{iks} can be ignored and the Hamiltonian (2.24) reduces to

$$H = \omega S^z + \sum_{ks} \omega_{ks} a_{ks}^+ a_{ks} + \sum_{ks} (g_{ks} a_{ks} S^+ + \text{H.C.}) , \tag{2.27}$$

where $S^{\pm}, S^z$ are the collective spin operators defined by

$$S^{\pm} = \sum_i S_i^{\pm}, \qquad S^z = \sum_i S_i^z. \tag{2.28}$$

In view of the commutation relations (2.11) one can easily show that $S^{\pm}, S^z$ satisfy the angular momentum commutation relations

$$[S^+, S^-] = 2S^z, \quad [S^z, S^+] = S^+, \quad [S^z, S^-] = -S^-, \quad [S^2, S] = 0 \tag{2.29}$$

i.e.

$$S \times S = iS.$$

Since S_i corresponds to spin-$\frac{1}{2}$ operators, it follows from the addition of the angular momenta that the operators S^2 and S^z have eigenvalues given by

$$S^2|S, m\rangle = S(S+1)|S, m\rangle, \qquad S^z|S, m\rangle = m|S, m\rangle,$$

where $m = -S, -S+1, ..., S-1, S$,

$$\begin{aligned} S &= 0, 1, 2, ... \tfrac{1}{2}N, \quad &\text{if } N \text{ is even}, \\ &= \tfrac{1}{2}, \tfrac{3}{2}, ... \tfrac{1}{2}N, \quad &\text{if } N \text{ is odd}. \end{aligned} \tag{2.30}$$

We will refer to the states $|S, m\rangle$ as the Dicke states [8]. These states in general are degenerate, the degeneracy is given by

$$\frac{N!(2S+1)}{(\tfrac{1}{2}N + S + 1)!\,(\tfrac{1}{2}N - S)!}. \tag{2.31}$$

The states corresponding to $S = \frac{1}{2}N$ are *not* degenerate. The state $|\frac{1}{2}N, m\rangle$ is the one in which $(\frac{1}{2}N + m)$ spins (atoms) are in the excited state and $(\frac{1}{2}N - m)$ spins in the ground state. The state $|\frac{1}{2}N, \frac{1}{2}N\rangle$ is given by

$$|\tfrac{1}{2}N, \tfrac{1}{2}N\rangle = \prod_{j=1}^{N} |1\rangle_j, \tag{2.32}$$

i.e. in this state all the atoms are in the excited state. The states $|\frac{1}{2}N, \frac{1}{2}N - n\rangle$ can be constructed from (2.32) by the application of the operator $(S^-)^n$ (apart from a normalization factor). The state

$$|\tfrac{1}{2}N - 1, \tfrac{1}{2}N - 1\rangle$$

can be constructed by using the properties of the permutation group. These are given by

$$|\tfrac{1}{2}N - 1, \tfrac{1}{2}N - 1; \alpha\rangle = N^{-1/2} \sum_{j=1}^{N} \exp\{(2\pi i\alpha/N)j\} \prod_{l \neq j} |2\rangle_j |1\rangle_l, \tag{2.33}$$

where $\alpha (= 1, 2, \ldots (N-1))$ gives the different degenerate states. The state $|\frac{1}{2}N - 2, \frac{1}{2}N - 2; \alpha\rangle$ can be constructed by forming the linear combination of the individual states $|1\rangle_j$, $|2\rangle_j$ in which two atoms are in the ground state and the rest in the excited state.

Finally, note that (2.24) can also be written in the form

$$H = \omega \sum_i S_i^z + \sum_{ks} \omega_{ks} a_{ks}^+ a_{ks} + \sum_{ks} (g_{ks} a_{ks} S_k^+ + \text{H.C.}), \tag{2.34}$$

where

$$S_k^{\pm} = \sum_i S_i^{\pm} \, e^{\pm i k \cdot r_i}, \quad S_k^z = \sum_i S_i^z . \tag{2.35}$$

For a *fixed* k the operators $S_k^{\pm}$, S_k^z still satisfy the angular momentum commutation relation, i.e.

$$S_k \times S_k = i S_k . \tag{2.36}$$

In the above we considered a collection of two-level atoms. One can similarly obtain the Hamiltonian for a multilevel atom interacting with a quantized radiation field. The dipole moment operator d in the present case is

$$d = \sum_{kl} d_{kl} A_{kl} , \tag{2.37}$$

where d_{kl} are the matrix elements and A_{kl} are given by

$$A_{kl} = |k\rangle \langle l| . \tag{2.38}$$

3. Weisskopf-Wigner Method

We initially discuss the very first method devised to treat spontaneous emission in the form in which it was originally presented by Weisskopf and Wigner. The wave function of the combined atomic + field system is written in the form

$$\psi = \Sigma b_n(t) \, \psi_n \, e^{-i E_n t} , \tag{3.1}$$

where the summation is only over those sets of states that can be reached by emitting the photon. E_n are the energy eigenvalues of the unperturbed states ψ_n. The amplitudes $b_n(t)$ obey the Schrödinger equation

$$i \dot{b}_m = \sum_n V_{mn} b_n \, e^{i \omega_{mn} t} , \tag{3.2}$$

where

$$\omega_{mn} = E_m - E_n \qquad (3.3)$$

and V is the interaction Hamiltonian. The method essentially consists of assuming for b_i (the suffix i indicating the initial excited atomic state with no photons present) an exponentially decaying solution

$$b_i(t) = e^{-\Gamma t}, \qquad (3.4)$$

and with a similar ansatz for $b_n(t)$. The Eq. (3.2) is then used to calculate the various damping coefficients.

To illustrate the method, we apply it to a study of the spontaneous emission from a single two-level atom. One makes the following ansatz for the wave function

$$\psi = b_1(t) \,|1, \{0\}\rangle \, e^{-iE_1 t} + \sum_k b_2^k(t) \,|2, \{k\}\rangle \, e^{-iE_2^k t}, \qquad (3.5)$$

where $|1\rangle$ and $|2\rangle$ are the excited and ground states of the atom, and $|\{0\}\rangle$ and $|\{k\}\rangle$ are respectively the vacuum state and the state of the radiation field, in which a photon in the mode k is present, where we have suppressed the polarization indices. The assumption (3.5) really means that the only possible transitions which the Hamiltonian allows are the *resonant* ones, i.e. the ones in which an atomic excitation is annihilated and a photon is created and vice versa. It should be noted that this is just the RWA which amounts to using the Hamiltonian (2.24). Therefore the RWA is implicit in the Weisskopf-Wigner theory. The Schrödinger Eq. (3.2) now leads to

$$i\dot{b}_1 = \sum_k V_{12}^k b_2^k \, e^{i(\omega - \omega_k)t}, \qquad (3.6a)$$

$$i\dot{b}_2^k = V_{12}^{k*} b_1 \, e^{-i(\omega - \omega_k)t}, \qquad (3.6b)$$

where ω, as before, is the energy separation between two atomic levels and $V_{12}^k \equiv \langle 1, \{0\}|V|2, \{k\}\rangle$. The initial conditions are

$$b_1(0) = 1, \quad b_2^k(0) = 0. \qquad (3.7)$$

We solve (3.6b) by using the ansatz (3.4) to obtain

$$-b_2^k(t) = V_{12}^{k*}(\omega_k - \omega + i\Gamma)^{-1} \left[\exp\{i(\omega_k - \omega)t - \Gamma t\} - 1\right]. \qquad (3.8)$$

On substituting (3.8) in (3.6a) we obtain an equation for Γ

$$-i\Gamma = \sum_k |V_{12}^k|^2 \, (\omega - \omega_k - i\Gamma)^{-1} \left[1 - \exp\{i(\omega - \omega_k)t + \Gamma t\}\right]. \qquad (3.9)$$

On taking the limit $L^3 \to \infty$, (3.9) reduces to

$$-i\Gamma = \int d\omega_k (\omega - \omega_k - i\Gamma)^{-1} \left[1 - \exp\{i(\omega - \omega_k)t + \Gamma t\}\right] \int d\Omega_k \varrho_k |V_{12}^k|^2, \qquad (3.10)$$

where $\varrho_k \, d\omega_k \, d\Omega_k$ represents the number of field modes in the frequency range ω_k and $\omega_k + d\omega_k$. For our ansatz to be consistent, the right-hand side of (3.10) should be time-independent. We are only interested in time intervals $t \gg 1/\omega$. Moreover, we will see *a posteriori* that $\Gamma \ll \omega$, then we can ignore Γ from the integrand. Under these conditions (3.10) reduces to

$$-i\Gamma \cong \int d\omega_k [-i\delta_+(\omega - \omega_k)] \int d\Omega_k \varrho_k |V_{12}^k|^2,$$

i.e.

$$\Gamma = \gamma + i\varDelta, \tag{3.11}$$

$$\gamma = \pi \int d\Omega_k \varrho_k(\omega_k) |V_{12}^k(\omega_k)|^2 \big|_{\omega_k = \omega},$$

$$\varDelta = P \int (\omega - \omega_k)^{-1} \, d\omega_k \int d\Omega_k \varrho_k |V_{12}^k|^2,$$

with P indicating the principal part of the integral. The real part of 2Γ is just equal to the transition probability per unit time that the system will make a transition from the excited state to the ground state, and the imaginary part of Γ gives the level shift of the excited state. Moreover, we see from (3.8) that

$$|b_2^k(\infty)|^2 = |V_{12}^k|^2 / \{(\omega_k - \omega - \varDelta)^2 + \gamma^2\} \equiv p_{ks}, \tag{3.12}$$

The distribution of the emitted photons is Lorentzian. On summing (3.12) over the two polarization directions and on integrating over a solid angle we find

$$p(\omega_k) \, d\omega_k = (\gamma \omega_k^2 / \pi \omega^2) [(\omega_k - \omega - \varDelta)^2 + \gamma^2]^{-1} \, d\omega_k. \tag{3.13a}$$

Therefore the distribution of energy is given by

$$I(\omega_k) \, d\omega_k = (\gamma / \pi \omega^2) \omega_k^3 \, d\omega_k / [(\omega_k - \omega - \varDelta)^2 + \gamma^2]. \tag{3.13}$$

If the $-A \cdot p$ interaction were used, then in place of (3.13) we would obtain

$$I(\omega_k) \, d\omega_k = (\gamma / \pi) \omega_k \, d\omega_k / [(\omega_k - \omega - \varDelta)^2 + \gamma^2], \tag{3.14}$$

which is in agreement with Weisskopf-Wigner [1] (see also Heitler [4]). We would adopt (3.13a) in view of the discussions of Lamb [17] and Power and Zienau [15]. It is also seen from (3.12) that only the *level shift* of the excited state appears; this is due to the fact that in the Hamiltonian we retained only the resonant terms. The shift of the ground state is due to the virtual transitions (nonresonant terms in the Hamiltonian) and hence does not appear in (3.12). We will discuss these terms in detail in Chapter 6 and Appendix A. In the rest of this section we ignore the level-shift terms.

As another illustration of the Weisskopf-Wigner method, we consider a more complicated system: the emission from a three-level atom with *nonequidistant spectrum*. Let the levels be arranged so that $E_1 > E_2 > E_3$. We assume that in the transition $E_1 \to E_2$ a photon in the mode k is emitted and that in the transition $E_2 \to E_3$ a photon in the mode l is emitted. Furthermore, the transition $E_1 \to E_3$ is forbidden, say from parity considerations. As before, we retain only the resonant transitions so that we express $|\psi\rangle$ in the form

$$|\psi\rangle = b_1(t)|1, \{0\}\rangle + \sum_k b_2^k(t)|2, \{k\}\rangle + \sum_{kl} b_3^{kl}(t)|3, \{k, l\}\rangle . \tag{3.15}$$

The amplitudes b now satisfy

$$i\dot{b}_1 = E_1 b_1 + \sum_k V_{12}^k b_2^k , \tag{3.16a}$$

$$i\dot{b}_2^k = (E_2 + \omega_k) b_2^k + V_{12}^{k*} b_1 + \sum_l V_{23}^{kl} b_3^{kl} , \tag{3.16b}$$

$$i\dot{b}_3^{kl} = (E_3 + \omega_k + \omega_l) b_3^{kl} + V_{23}^{kl*} b_2^k , \tag{3.16c}$$

where

$$V_{12}^k \equiv \langle 1, \{0\}|V|2, \{k\}\rangle, \quad V_{23}^{kl} \equiv \langle 2, \{k\}|V|3, \{k, l\}\rangle . \tag{3.17}$$

We now make the ansatz

$$b_1(t) = e^{-\gamma_1 t}, \quad b_2^k(t) = A_k \{e^{-\gamma_1 t} - e^{-\gamma_2^k t}\} , \tag{3.18a}$$

$$b_3^{kl}(t) = B_{kl} e^{-\gamma_1 t} + C_{kl} e^{-\gamma_2^k t} - (B_{kl} + C_{kl}) e^{-\gamma_3^{kl} t} , \tag{3.18b}$$

which is consistent with the initial condition $b_1(0) = 1$, $b_2^k(0) = b_3^{kl}(0) = 0$. From (3.16c) and (3.18) we obtain the relations

$$B_{kl} = -V_{23}^{kl*} A_k / \{i\gamma_1 + (E_3 + \omega_k + \omega_l)\} , \tag{3.19}$$

$$C_{kl} = V_{23}^{kl*} A_k / \{i\gamma_2^k + (E_3 + \omega_k + \omega_l)\} , \quad \gamma_3^{kl} = i(E_3 + \omega_k + \omega_l) .$$

On substituting (3.18), (3.19) in (3.16b) we find that

$$\mathrm{Im}\,\gamma_2^k = E_2 + \omega_k, \quad \mathrm{Re}\,\gamma_2^k = \gamma_{32} , \tag{3.20a}$$

$$A_k = -V_{12}^{k*} \left\{ i\gamma_1 + E_2 + \omega_k - \frac{i\gamma_{32}}{(E_3 - E_2)} (E_3 + \omega_k - \mathrm{Im}\,\gamma_1) \right\}^{-1} . \tag{3.20b}$$

In obtaining (3.20) we ignored the small level-shift terms and made the approximations in the way that we did on passing from (3.10) to (3.11). Similarly, on substituting (3.20) in (3.16a) we obtain

$$\gamma_1 = \gamma_{21} + iE_1 . \tag{3.21}$$

In (3.20a) and (3.21) $2\gamma_{ij}$ is the transition probability per unit time for a transition from the state $|j\rangle$ to $|i\rangle$. We have now obtained all the amplitudes for calculating transition rates. We have, for instance

$$b_3^{kl}(\infty) = -(B_{kl} + C_{kl})\, e^{-i(E_3 + \omega_k + \omega_l)t}$$

$$= -V_{23}^{kl*}V_{12}^{k*}\{\omega_k + E_2 - E_1 + i(\gamma_{21} - \gamma_{32})\}\,\{i\gamma_{32} - E_2 + E_3 + \omega_l\}^{-1}$$

$$\cdot\{i\gamma_{21} - E_1 + E_3 + \omega_k + \omega_l\}^{-1}$$

$$\cdot\left\{i\gamma_{21} - E_1 + E_2 + \omega_k - \frac{i\gamma_{32}}{(E_3 - E_2)}(E_3 + \omega_k - E_1)\right\}^{-1}$$

$$\cdot e^{-i(E_3 + \omega_k + \omega_l)t}$$

$$\approx -V_{23}^{kl*}V_{12}^{k*}\{i\gamma_{32} - \omega_{23} + \omega_l\}^{-1}\{i\gamma_{21} - \omega_{13} + \omega_k + \omega_l\}^{-1}$$

$$\cdot e^{-i(E_3 + \omega_k + \omega_l)t}, \tag{3.22}$$

where

$$\omega_{ij} = E_i - E_j. \tag{3.23}$$

(3.22) gives the distribution of photons in the modes k and l:

$$|b_3^{kl}|^2 = |V_{23}^{kl}|^2\,|V_{12}^{k}|^2\,\{\gamma_{32}^2 + (\omega_{23} - \omega_l)^2\}^{-1}\,\{\gamma_{21}^2 + (\omega_{13} - \omega_k - \omega_l)^2\}^{-1}. \tag{3.24}$$

The individual distributions p_{ks}, $p_{l\lambda}$ are obtained by

$$p_{ks} = \sum_l |b_3^{ks,l\lambda}|^2, \qquad p_{l\lambda} = \sum_{ks} |b_3^{ks,l,\lambda}|^2, \tag{3.25}$$

where the summations are also over the two polarization directions. Therefore we have for individual distributions

$$p_{ks} = |V_{12}^{ks}|^2\,\gamma_{21}^{-1}(\gamma_{23} + \gamma_{12})\{(\omega_{ks} - \omega_{12})^2 + (\gamma_{32} + \gamma_{12})^2\}^{-1}, \tag{3.26a}$$

$$p_{l\lambda} = |V_{23}^{l\lambda}|^2\,\{(\omega_{l\lambda} - \omega_{23})^2 + \gamma_{32}^2\}^{-1}, \tag{3.26b}$$

which show that the distribution of the photon emitted in the transition $|2\rangle \to |3\rangle$ is independent of the presence of the first level; however, the linewidth in the transition $|1\rangle \to |2\rangle$ is the sum of the widths of the two levels.

The fact that the linewidth is the sum of the widths of the two levels applies only to a system with nonequidistant spectrum. If we were to consider the emission from a harmonic oscillator (which has an equidistant spectrum) then the above analysis, if correct, would give a linewidth proportional to $(2n - 1)$, where n is the excitation of the oscillator. However, it is not true, for there is no unique way of determining which photon came as a result of which transition. We will treat the harmonic oscillator model in Chapter 12 by means of master equation techniques.

4. Heitler-Ma Method

We next discuss the Heitler-Ma method, which is extensively used in atomic physics; our discussion is brief as the method is described in great detail in Heitler's book [4].

The wave function for the combined atom–field system is written in the form (3.1) with the amplitude coefficients $b_n(t)$ satisfying (3.2) and the initial condition

$$b_n(0) = \delta_{ni} , \tag{4.1}$$

where i indicates the initial state. The initial condition (4.1) can be built in to (3.2) by modifying it to

$$\dot{b}_n(t) = - i \Sigma V_{nm} b_m e^{i(E_n - E_m)t} + \delta_{ni} \delta(t) . \tag{4.2}$$

These equations are further assumed to hold for all t by requiring that $b_n(t) = 0$ for $t < 0$. On introducing the Fourier transforms defined by

$$b_n(t) = - (1/2\pi i) \int\limits_{-\infty}^{+\infty} dE\, G_{ni}(E)\, e^{i(E_n - E)t} , \tag{4.3}$$

(4.2) reduces to

$$(E - E_n)\, G_{ni}(E) = \sum_m V_{nm} G_{mi}(E) + \delta_{ni} . \tag{4.4}$$

To solve (4.4), we introduce

$$G_{ni}(E) = - i U_{ni}(E)\, G_{ii}(E)\, \delta_+(E - E_n), \quad n \neq i . \tag{4.5}$$

On substituting (4.5) in (4.4) we obtain the integral equation for U:

$$U_{ni}(E) = V_{ni} - i \sum_{m \neq i} V_{nm} U_{mi}(E)\, \delta_+(E - E_m), \quad n \neq i . \tag{4.6}$$

From (4.4) we find the $G_{ii}(E)$ is given by

$$G_{ii}(E) = \{E - E_i - \Sigma_{ii}(E)\}^{-1} , \tag{4.7}$$

where

$$\Sigma_{ii}(E) = V_{ii} - i \sum_{m \neq i} V_{im} U_{mi}(E)\, \delta_+(E - E_m) , \tag{4.8}$$

and therefore from (4.3) we find that

$$b_i(t) = - (1/2\pi i) \int\limits_{-\infty}^{+\infty} dE\, e^{i(E_i - E)t} \{E - E_i - \Sigma_{ii}(E)\}^{-1} , \tag{4.9}$$

$$b_n(t) = - (1/2\pi i) \int\limits_{-\infty}^{+\infty} dE\, e^{i(E_n - E)t} \{ - i \delta_+(E - E_n)\} \tag{4.10}$$
$$\cdot U_{ni}(E)\, \{E - E_i - \Sigma_{ii}(E)\}^{-1} .$$

In view of (4.1) we have from (4.10)

$$0 = -(1/2\pi i) \int_{-\infty}^{+\infty} dE \{-i\delta_+(E - E_n)\} U_{ni}(E) \{E - E_i - \Sigma_{ii}(E)\}^{-1}, \quad (4.11)$$

and hence $b_n(t)$ can also be written in the form

$$b_n(t) = -(1/2\pi i) \int_{-\infty}^{+\infty} dE\, U_{ni}(E) \{E - E_i - \Sigma_{ii}(E)\}^{-1} (E - E_n)^{-1} \quad (4.12)$$
$$\cdot \{e^{i(E_n - E)t} - 1\},$$

which clearly exhibits the initial condition. It is obvious from (4.12) that

$$b_n(\infty) = U_{ni}(E_n)/\{E_n - E_i - \Sigma_{ii}(E_n)\}, \quad (4.13)$$

and hence the probability distribution is

$$|b_n(\infty)|^2 = |U_{ni}(E_n)|^2/\{(E_n - E_i - \Delta_n)^2 + \gamma_n^2\}, \quad (4.14)$$

where

$$\Delta_n = \mathrm{Re}\,\Sigma_{ii}(E_n), \quad \gamma_n = -\mathrm{Im}\,\Sigma_{ii}(E_n). \quad (4.15)$$

It is therefore clear that the real part of $\Sigma_{ii}(E)$ as given by (4.8) represents the level shift and the imaginary part the level width. If the physical problem is simple enough, such as the decay of a two-level atom, the frequency distribution of the final state would be Lorentzian; otherwise it may be very different because of the strong dependence of U_{ni} upon E_i and E_n. We will later see examples of more complicated behavior. Moreover, in the above formula only the level displacement of the initial state appears and hence is asymmetric with respect to the initial and final states.

$\Sigma_{ii}(E)$ to lowest order in the interaction between the atom and field is given by

$$\Sigma_{ii}(E) \approx \sum_{m \neq i} \{-i\delta_+(E - E_m)\} V_{im} V_{mi} = \Delta - i\gamma, \quad (4.16)$$

$$\gamma = \sum_{m \neq i} \pi\delta(E - E_m) V_{im} V_{mi}, \quad (4.17)$$
$$\Delta = P \sum_{m \neq i} V_{im} V_{mi}(E - E_m)^{-1},$$

where we have assumed that $V_{ii} = 0$. If the energy dependence of γ and Δ is ignored, then it follows from (4.19) that

$$b_i(t) = \exp\{-i\Delta t - \gamma t\}, \quad (4.18)$$

which represents the familiar exponential decay which Weisskopf and Wigner took as their starting point. It can be further seen from the second-order perturbation theory that 2γ represents the total transition probability per unit time from the state $|i\rangle$ to all other states.

It is apparent that the Heitler-Ma method provides us with a systematic way of calculating level shifts and level widths to all orders in the coupling coefficient between the field and the matter. Moreover, it derives the results of the exponential decay theory, whereas in Weisskopf-Wigner theory exponential decay was justified *a posteriori*.

As an application of the above formalism, let us consider spontaneous emission from a system of two atoms, each of which is assumed to have two levels, prepared initially with one in the excited state and the other in the ground state, i.e.

$$|i\rangle = |1\rangle_1 |2\rangle_2 |\{0\}\rangle . \tag{4.19}$$

We consider only the resonant transitions, so that the intermediate and final states are

$$|\mu\rangle = |2\rangle_1 |1\rangle_2 |\{0\}\rangle , \tag{4.20a}$$

$$|f\rangle = |2\rangle_1 |2\rangle_2 |\{k\}\rangle , \tag{4.20b}$$

i.e. in the intermediate state the atomic excitation is transferred to the second atom, the field remaining in the vacuum state, and in the final state both atoms are left in the ground state and a photon is emitted. To obtain the distribution of the emitted photon one should calculate $|b_f(\infty)|^2$, which is determined from the knowledge of U_{fi}, Σ_{ii} (cf. Eq. (4.14)). From (4.6) we have

$$U_{fi}(E) = V_{fi} + V_{f\mu} U_{\mu i}(E) \{-i\delta_+(E - E_\mu)\} ,$$

$$U_{\mu i}(E) = V_{\mu i} - i \sum_k V_{\mu f} U_{fi}(E) \delta_+(E - E_f) . \tag{4.21}$$

Solving (4.21) for $U_{fi}(E)$, we obtain

$$U_{fi}(E) = V_{fi} + V_{f\mu}(V_{\mu i} + \Sigma_{\mu i})(E - E_\mu - \Sigma_{\mu\mu})^{-1} , \tag{4.22}$$

$$\Sigma_{\mu\mu} = -i \sum_k |V_{\mu f}|^2 \delta_+(E - E_f), \qquad \Sigma_{\mu i} = -i \sum_k V_{\mu f} V_{fi} \delta_+(E - E_f) . \tag{4.23}$$

We further need to calculate $\Sigma_{ii}(E)$, which from (4.8) is equal to

$$\Sigma_{ii}(E) = V_{ii} - i V_{i\mu} U_{\mu i}(E) \delta_+(E - E_\mu) - i \sum_k V_{if} U_{fi}(E) \delta_+(E - E_f)$$

$$= \frac{(V_{\mu i} + \Sigma_{\mu i})}{(E - E_\mu - \Sigma_{\mu\mu})}(V_{i\mu} + \Sigma_{i\mu}) - i \sum_k |V_{if}|^2 \delta_+(E - E_f)$$

$$= \Sigma'_{ii}(E) + (E - E_\mu - \Sigma_{\mu\mu})^{-1} (V_{\mu i} + \Sigma_{\mu i})(V_{i\mu} + \Sigma_{i\mu}) , \tag{4.24}$$

where $\Sigma'_{ii}(E)$ is defined by (4.23) with μ replaced by i. The probability distribution is given by

$$b_f(\infty) = U_{fi}(E_f)/\{E_f - E_i - \Sigma_{ii}(E_f)\} \,. \tag{4.25}$$

The numerical curves for the line shape, as obtained from (4.25), are reported in Czarnik and Fontana [18]. It is easily seen that $\Sigma_{\mu\mu}$ and Σ_{ii} give respectively the level shift and the level width of the intermediate and the initial state. Moreover, $\Sigma_{\mu i}$ is the one which depends on the distance between the two atoms, the real and imaginary parts being related to γ_{ij} and Ω_{ij} of Chapter 6. For other examples treated by the present method, we refer to [19] and the literature references cited therein.

5. Goldberger-Watson Method

In this section we review the Goldberger-Watson approach to spontaneous emission. These authors developed very elegant projection operator techniques which they applied to the wave function of the system. Their methods are closely related to the master-equation methods that form the central theme of this article. The major difference is that in the master-equation framework one works with the density operator and this allows the treatment of more general states. Nevertheless, the Goldberger-Watson method has proved quite useful in discussing certain aspects of radiative decay theory.

We start from the Schrödinger equation

$$i\,\partial\psi/\partial t = H\psi \,, \tag{5.1}$$

and take the Laplace transform

$$z\hat{\psi}(z) - \psi(0) = -iH\hat{\psi}(z) \,, \tag{5.2}$$

where the Laplace-transformed expressions are defined by

$$\hat{\psi}(z) = \int_0^\infty e^{-zt}\psi(t)\,dt, \quad \mathrm{Re}\,z > 0 \,. \tag{5.3}$$

We introduce the projection operator $\mathscr{P}$ having the property $\mathscr{P}^2 = \mathscr{P}$, the explicit form of which will depend on the physical problem under consideration. On applying $\mathscr{P}$ and $(1 - \mathscr{P})$ to both sides of (5.2) we obtain

$$z\mathscr{P}\hat{\psi} - \mathscr{P}\psi(0) = -i\{\mathscr{P}H\mathscr{P}\hat{\psi} + \mathscr{P}H(1 - \mathscr{P})\,\hat{\psi}\} \,, \tag{5.4a}$$

$$z(1 - \mathscr{P})\,\hat{\psi} - (1 - \mathscr{P})\,\psi(0) = -i\{(1 - \mathscr{P})\,H\mathscr{P}\hat{\psi} + (1 - \mathscr{P})\,H(1 - \mathscr{P})\,\hat{\psi}\} \,. \tag{5.4b}$$

On solving (5.4b) we find

$$(1 - \mathscr{P}) \hat{\psi} = \{z + i(1 - \mathscr{P}) H (1 - \mathscr{P})\}^{-1}$$
$$\cdot \{(1 - \mathscr{P}) \psi(0) - i(1 - \mathscr{P}) H \mathscr{P} \hat{\psi}\}, \tag{5.5}$$

which on substituting in (5.4a) leads to

$$\mathscr{P} \hat{\psi} = \{z + i\mathscr{P} H \mathscr{P} + \mathscr{P} H [z + i(1 - \mathscr{P}) H (1 - \mathscr{P})]^{-1} (1 - \mathscr{P}) H\}^{-1}$$
$$\cdot \{\mathscr{P} \psi(0) - i\mathscr{P} H [z + i(1 - \mathscr{P}) H (1 - \mathscr{P})]^{-1} (1 - \mathscr{P}) \psi(0)\}. \tag{5.6}$$

We will choose the projection operator so that $\mathscr{P}\psi$ is the relevant part of the wave function. To see how (5.6) can be used, we apply it to study the spontaneous emission from a single atom. We choose

$$\mathscr{P} = |i\rangle \langle i|, \tag{5.7}$$

where $|i\rangle$ is the initial state of the atom + field system. The field is, of course, in the vacuum state. It is obvious that

$$\mathscr{P}\psi(0) = \psi(0), \quad (1 - \mathscr{P}) \psi(0) = 0. \tag{5.8}$$

On taking the matrix element of (5.6) with $|\psi\rangle$ we obtain

$$\langle i|\hat{\psi}\rangle = \langle i|[z + i\mathscr{P} H \mathscr{P} + \mathscr{P} H \{z + i(1 - \mathscr{P}) H (1 - \mathscr{P})\}^{-1} (1 - \mathscr{P}) H]^{-1} |i\rangle$$
$$= [z + iE_i + \Sigma_{ii}(z)] \equiv \mathscr{S}_{ii}(z), \tag{5.9}$$

where

$$\Sigma_{ii}(z) = \langle i|\mathscr{P} H \{z + i(1 - \mathscr{P}) H (1 - \mathscr{P})\}^{-1} (1 - \mathscr{P}) H |i\rangle$$
$$= \langle i| V \{z + i(1 - \mathscr{P}) H (1 - \mathscr{P})\}^{-1} (1 - \mathscr{P}) V |i\rangle. \tag{5.10}$$

In the above we assumed that $\langle i|V|i\rangle = 0$. It should be noted that $\langle i|\hat{\psi}\rangle$ represents the Laplace transform of the probability amplitude that the system remains in its initial state. (5.9) should be compared with (4.9). The operator $\Sigma(z)$ is the so called self-energy operator and contains the effect of the interaction to all orders. In terms of the eigenstates ψ_p of the operator $(1 - \mathscr{P}) H (1 - \mathscr{P})$, (5.10) can be expressed as

$$\Sigma_{ii}(z) = \sum_p \langle i| V (1 - \mathscr{P}) |\psi_p\rangle \langle \psi_p| (1 - \mathscr{P}) V |i\rangle (z + iE_p)^{-1}. \tag{5.11}$$

It is evident from (5.11) that the sign of the imaginary part of $\Sigma_{ii}(z)$ is opposite to the sign of the imaginary part of z, which would in general imply a decay of the probability amplitude. To illustrate it, we consider Σ_{ii} to second order in the coupling coefficient

$$\Sigma_{ii}(z) \approx \sum_\mu \langle i| V |\mu\rangle \langle \mu| V |i\rangle (z + iE_\mu)^{-1}, \tag{5.12}$$

where $|\mu\rangle$ and E_μ are the eigenstates and the eigenvalues of the unperturbed Hamiltonian. The sum is over all possible states that can be reached via one photon transition, the photon can be emitted into any mode and hence the above summation also includes an integration over the continuum of modes. The probability amplitude that the system (atom) will be found in the excited state is

$$\mathscr{S}_{ii}(t) = (1/2\pi i)\oint dz\, e^{zt}[z + iE_i + \sum_\mu |V_{i\mu}|^2 (z + iE_\mu)^{-1}]^{-1}, \tag{5.13}$$

where the integration is along any line parallel to the imaginary axis in the right half-plane such that no singularities of $(z + iH)^{-1}$ lie to the right of it. For time intervals greater than "characteristic times" (5.13) can be approximated by

$$\mathscr{S}_{ii}(t) = \exp\{-i(E_i + \Delta)t - \gamma t\}, \tag{5.14}$$

where

$$\Delta = \mathrm{Im}\,\Sigma_{ii}(-iE_i), \quad \gamma = \mathrm{Re}\,\Sigma_{ii}(-iE_i). \tag{5.15}$$

Thus, as before, the real and imaginary parts of Σ_{ii} give rise to the level width and level shift, respectively. It should be noted that by ignoring the variation of $\Sigma_{ii}(z)$ in the integrand in (5.13), i.e. by replacing z by $-iE_i$, an error has been introduced and hence for extremely large times $\mathscr{S}_{ii}(t)$ need not decay exponentially [cf. Ref. [5], p. 450, Eq. (116)].

We can now calculate the probability that the system decays to the lower state for the case of a two-level atom with the emission of a photon in the mode k. This probability amplitude $\mathscr{S}_{2k,1}$ is given by

$$\begin{aligned}
\hat{\mathscr{S}}_{2k,1} &= \langle 2, \{k\}|(z + iH)^{-1}|1, \{0\}\rangle \\
&= \langle 2, \{k\}|(z + iH_0)^{-1} + (z + iH_0)^{-1}V(z + iH)^{-1}|1, \{0\}\rangle,
\end{aligned} \tag{5.16}$$

$$= (z + iE_2 + iE_k)^{-1}\langle 2, \{k\}|V(1 - \mathscr{P} + \mathscr{P})(z + iH)^{-1}|1, \{0\}\rangle$$

$$\approx (z + iE_2 + iE_k)^{-1} V_{1,2}^{k*}\hat{\mathscr{S}}_{11}, \tag{5.17}$$

where in going from (5.16) to (5.17) we have ignored the contribution of the term $(1 - \mathscr{P})$ as it is of higher order in perturbation. On combining (5.14) and (5.17), we obtain

$$|\mathscr{S}_{2k_1}(t)|^2 = |V_{12}^k|^2 (1 + e^{-2\gamma t} - 2e^{-\gamma t}\cos Xt)(\gamma^2 + X^2)^{-1}$$

$$X = (\omega_k - E_1 + E_2 - \Delta), \tag{5.18}$$

and hence the line shape is Lorentzian. Again the level shift of the lower level is missing from (5.18), due to the fact that ground state has been inadequately treated, for one ought to take into account the virtual transitions. We will not discuss such effects within the present formalism and we refer to the work of Kroll [20]. We will, however, take up such effects in connection with master equations.

The generalizations of the above formalism have been considered by Goldberger, Goldhaber and Watson [21], and Mower [22]. Here we discuss briefly the sequential decay theory due to Mower and apply it to a three-level atom and a system of two two-level atoms. We slightly recast the formalism of the first part of this section: the time evolution operator $U(t)$ is

$$U(t) = e^{-iHt}, \tag{5.19}$$

and its one-sided Fourier transform $R(z)$ is given by

$$R(z) = -i \int e^{-iHt} e^{izt} dt = (z - H)^{-1}. \tag{5.20}$$

The relation inverse to (5.20) is

$$U(t) = (1/2\pi i) \oint dz \, e^{-izt} R(z), \tag{5.21}$$

where the contour runs from $+\infty$ to $-\infty$ above the singularities of $R(z)$ on the real axis. On applying the projection operators $\mathscr{P}$ and $(1 - \mathscr{P})$ to the equation

$$(z - H) R(z) = 1, \tag{5.22}$$

we obtain the equations

$$\mathscr{P}(z - H) \mathscr{P} R(z) \mathscr{P} + \mathscr{P}(z - H)(1 - \mathscr{P}) R(z) \mathscr{P} = \mathscr{P}, \tag{5.23a}$$

$$(1 - \mathscr{P})(z - H) \mathscr{P} R(z) \mathscr{P} + (1 - \mathscr{P})(z - H)(1 - \mathscr{P}) R(z) \mathscr{P} = 0. \tag{5.23b}$$

From (5.23b) we have

$$(1 - \mathscr{P}) R(z) \mathscr{P} = \{z - (1 - \mathscr{P}) H(1 - \mathscr{P})\}^{-1} (1 - \mathscr{P}) H \mathscr{P} R(z) \mathscr{P}, \tag{5.23c}$$

and then from (5.23a)

$$\mathscr{P} R(z) \mathscr{P} = \mathscr{P} \{z - H_0 - \mathscr{P} \Sigma(z) \mathscr{P}\}^{-1}, \tag{5.24}$$

where the self-energy operator $\Sigma(z)$ is given by

$$\Sigma(z) = V + V(1 - \mathscr{P}) \{z - (1 - \mathscr{P}) H(1 - \mathscr{P})\}^{-1} (1 - \mathscr{P}) V. \tag{5.25}$$

From (5.25) it is easy to show

$$(1 - \mathscr{P}) \{z - (1 - \mathscr{P}) H(1 - \mathscr{P})\}^{-1} (1 - \mathscr{P}) V$$
$$= (1 - \mathscr{P})(z - H_0)^{-1} (1 - \mathscr{P}) \Sigma(z),$$

and we may write (5.23c) as

$$(1 - \mathscr{P}) R(z) \mathscr{P} = (1 - \mathscr{P})(z - H_0)^{-1} (1 - \mathscr{P}) \Sigma(z) \mathscr{P} R(z) \mathscr{P}. \tag{5.26}$$

If for $\mathscr{P}$ we take the operator (5.7), then $\mathscr{P}R(z)\mathscr{P}$ will give the probability amplitude that the system will remain in its initial state. The matrix element of (5.26) with $\langle 2, \{k\}| \dots |1, \{0\}\rangle$ gives the probability amplitude that the system will make a transition to the ground state by emitting a photon. The results (5.17) and (5.13) can be obtained by replacing $\Sigma(z)$ in (5.26) by V and in (5.24) by calculating it to second order in the interaction V.

The analytic properties of the operator $\Sigma(z)$ are discussed at length by Goldberger and Watson. They find that $\Sigma(z)$ is analytic everywhere except on the real axis where it has a number of discrete poles super-imposed onto a continuum. Under the condition

$$\langle i| V(1-\mathscr{P})V|i\rangle < \infty$$

they find that $\Sigma_{ii}(z)$ can be written in the form

$$\Sigma_{ii}^{\pm}(z) = D_{ii}(z) \mp iI_{ii}(z), \tag{5.27}$$

where the superscripts $\pm$ refer respectively to the upper and lower region of the complex z plane and D is the real part of Σ. D and I are related by the dispersion relation

$$D(z) = V_{ii} - \pi^{-1}\,\mathrm{P}\int\limits_{-\infty}^{+\infty} I(x')\,\mathrm{d}x'/(x'-z),$$

under the condition

$$\lim_{|\varepsilon|\to\infty}\{\Sigma_{ii}(z) - V_{ii}\} = O(\varepsilon^{-q}), \quad q > 0.$$

Moreover, $I_{ii}(z)$ for real z is different from zero only for $z > z_0$, where z_0 is the lowest eigenvalue in the continuum. To compute the decay proba-bilities it is necessary to continue analytically the function $\Sigma_{ii}(z)$ onto the second Riemann sheet, denoted by $\Sigma_{ii}^{\mathrm{II}}(z)$, and this continuation is carried on by defining for real x

$$\Sigma_{ii}^{\mathrm{II}}(x - i\eta) = \Sigma_{ii}^{\mathrm{I}}(x + i\eta), \quad \eta > 0. \tag{5.28}$$

We now modify the above equations so as to make them applicable to sequential decay such as occurs in case of emission from a collection of atoms or in case of cascade transitions. We first discuss the case when only one intermediate state is available, and we introduce a projection operator $\mathscr{P}_j$ corresponding to it, also the related projection operators

$$Q = 1 - \mathscr{P}, \quad Q_j = 1 - \mathscr{P} - \mathscr{P}_j. \tag{5.29}$$

To make it evident that the system goes through an intermediate state, we factorize $\Sigma(z)$ by considering the projection of the operator $Q[z - QHQ]^{-1}$. We use the same procedure as we used in conjunction with (5.22), i.e. we multiply on the right by $\mathscr{P}_j$ and on the left by $\mathscr{P}_j$ and Q_j. A simple analysis shows that

$$
\begin{aligned}
\Sigma(z) &= \Sigma^{(j)}(z)\left[1 + \mathscr{P}_j G^{(j)}(z)\,\mathscr{P}_j \Sigma^{(j)}(z)\right],\\
\Sigma^{(j)}(z) &= V + V Q_j (z - Q_j H Q_j)^{-1} Q_j V,\\
\mathscr{P}_j G^{(j)}(z)\,\mathscr{P}_j &= \mathscr{P}_j\{z - \mathscr{P}_j H_0 - \mathscr{P}_j \Sigma^{(1)}(z)\,\mathscr{P}_j\}^{-1}.
\end{aligned}
\tag{5.30}
$$

In the present case we are interested in the projection $Q_j R(z)\,\mathscr{P}$ which is equal to

$$
\begin{aligned}
Q_j R(z)\,\mathscr{P} &= Q_j (z - H_0)^{-1} Q_j \Sigma(z)\,\mathscr{P} R(z)\,\mathscr{P}\\
&= Q_j (z - H_0)^{-1} Q_j \Sigma^{(j)}(z)\left[1 + \mathscr{P}_j G^{(j)}(z)\,\mathscr{P}_j \Sigma^{(j)}(z)\right]\mathscr{P} R(z)\,\mathscr{P}.
\end{aligned}
\tag{5.31}
$$

The transition amplitude between the initial state $|i\rangle$ and the final state $|f\rangle$ is obtained by taking the matrix element of (5.31) and is for $t \to \infty$

$$
\mathscr{S}_{fi} = e^{-iE_f t}\langle f|\,\Sigma^{(j)}(E_f)\left[1 + \mathscr{P}_j G^{(j)}(E_f)\,\mathscr{P}_j \Sigma^{(j)}(E_f)\right]|i\rangle\,\mathscr{S}_{ii}(E_f).
\tag{5.32}
$$

As an application of (5.32) let us consider the case of a three-level atom initially in its uppermost state. The atom decays to its ground state with the emission of two photons. The appropriate projection operators for this problem are

$$
\mathscr{P} = |1\rangle\langle 1|,\qquad \mathscr{P}_j = |2,\{k\}\rangle\langle 2,\{k\}|,
\tag{5.33}
$$

the final state being $|3,\{k,l\}\rangle$. The transition amplitude is approximately equal to

$$
\begin{aligned}
\mathscr{S}_{3\{k,l\};1\{0\}} &\approx \exp\{-iEt\}\,\langle 3\,\{k,l\}|\,V\,|2,\{k\}\rangle\langle 2,\{k\}|\\
&\qquad \cdot G^{(j)}(E)\,|2,\{k\}\rangle\langle 2,\{k\}|\,V\,|1\,\{0\}\rangle\,\mathscr{S}_{ii}(E)\\
&\approx e^{-iEt} V_{23}^{kl*} V_{12}^{k*}\left[z - E_1 - \Sigma_{11}(E)\right]^{-1}\left[z - E_2 - E_k - \Sigma_{22}(E_2)\right]^{-1},
\end{aligned}
$$

$$
E \equiv E_3 + \omega_k + \omega_l,
$$

and the probability is

$$
\begin{aligned}
|\mathscr{S}_{3\{kl\},1\{0\}}|^2 &= |V_{23}^{kl}|^2\,|V_{12}^{k}|^2\left[(\omega_k + \omega_l - \omega_{13}')^2 + \gamma_{21}^2\right]^{-1}\\
&\qquad \cdot\left[(\omega_l - \omega_{23}')^2 + \gamma_{32}^2\right]^{-1},
\end{aligned}
\tag{5.34}
$$

where

$$
\begin{aligned}
\omega_{13}' &= \omega_{12}' + \omega_{23}',\qquad \omega_{12}' = E_1 + \operatorname{Re}\Sigma_{11} - E_2 - \operatorname{Re}\Sigma_{22},\\
\omega_{23}' &= E_2 + \operatorname{Re}\Sigma_{22} - E_3,\qquad \gamma_{21} = \operatorname{Im}\Sigma_{11},\qquad \gamma_{32} = \operatorname{Im}\Sigma_{22}.
\end{aligned}
\tag{5.35}
$$

(5.34) is the standard result [cf. our Eq. (3.24)]. Again as before, the level shift of the ground state is missing. The results for the individual frequency distributions are obtained by integrating one over the other.

As another example let us consider a three-level atom (e.g. $2S$, $2P$, $1S$ states in atomic hydrogen) in which the first two levels are coupled by an external perturbation H_{ext} and the second level decays spontaneously to the ground state (third level, say, $1S$) emitting a photon. Thus, the levels $|2\rangle$ and $|3\rangle$ are radiatively coupled whereas $|1\rangle$ and $|2\rangle$ are coupled by H_{ext}. Such a system is of interest in connection with Lamb shift measurements. In this case the appropriate projection operators are

$$\mathscr{P} = |1, \{0\}\rangle \langle 1, \{0\}|, \quad \mathscr{P}_j = |2, \{0\}\rangle \langle 2, \{0\}|. \tag{5.36}$$

Note that $(1 - \mathscr{P}) H_{ext} (1 - \mathscr{P}) = 0$. The probability amplitude for the initial state is

$$\mathscr{S}_{11}(z) = (z - E_1 - \Sigma_{11}(z))^{-1}, \tag{5.37}$$

where

$$\begin{aligned}
\Sigma_{11}(z) &= \langle 1, \{0\}| (V + H_{ext}) + (V + H_{ext}) Q [z - Q(H_{ext} + V) Q]^{-1} \\
&\quad \cdot Q(H_{ext} + V)|1, \{0\}\rangle \\
&= |\langle 1, \{0\}| H_{ext} |2, \{0\}\rangle|^2 \langle 2, \{0\} |[z - H_0 - QVQ]^{-1}|2, \{0\}\rangle \\
&\approx |\langle 1, \{0\}| H_{ext} |2, \{0\}\rangle|^2 (z - E_2 + i\gamma_2)^{-1},
\end{aligned} \tag{5.38}$$

where γ_2 is the damping coefficient associated with the level $|2\rangle$, and we have ignored its level shift. The probability amplitude for the intermediate state is

$$\begin{aligned}
\mathscr{S}_{21}(z) &= \mathscr{S}_{11}(z) \langle 2, \{0\} | H_{ext}|1, \{0\}\rangle \\
&\quad \cdot \langle 2, \{0\} |[z - H_0 - Q(H_{ext} + V) Q]^{-1}|2, \{0\}\rangle \\
&\approx \mathscr{S}_{11}(z) \langle 2, \{0\} | H_{ext}|1, \{0\}\rangle [z - E_2 + i\gamma_2]^{-1}.
\end{aligned} \tag{5.39}$$

The probability amplitude $\mathscr{S}_{fi}(z)$ for the final state $|f\rangle \equiv |3, \{k\}\rangle$ from (5.32) is given by

$$\begin{aligned}
\mathscr{S}_{f1}(z) &\approx \mathscr{S}_{11}(z) (z - E_f)^{-1} \langle f | V |2, \{0\}\rangle \langle 2, \{0\} |[z - H_0 - \mathscr{P}_j \Sigma^{(j)}(z) \\
&\quad \cdot \mathscr{P}_j]^{-1}|2, \{0\}\rangle \langle 2, \{0\}| H_{ext}|1, \{0\}\rangle \\
&\approx \mathscr{S}_{11}(z) (z - E_f)^{-1} \langle f | V |2, \{0\}\rangle \langle 2, \{0\}| H_{ext}|1, \{0\}\rangle (z - E_2 + i\gamma_2)^{-1}.
\end{aligned} \tag{5.40}$$

The results (5.37) to (5.40) are identical to those obtained by the Heitler-Ma method in Ref. [19]. We will again discuss this system using master equations in Chapter 18.

In the derivation leading to (5.32) it was assumed that only one intermediate level was available. For the case when n intermediate levels are available, the formalism is easily generalized by the introduction of new projection operators defined by

$$Q_m = 1 - \mathscr{P} - \sum_{j=1}^{m} \mathscr{P}_j, \quad m = 1, 2, \ldots, n. \tag{5.41}$$

The generalization of (5.31) is now

$$Q_n R(z) \mathscr{P}$$
$$= Q_n (z - H_0)^{-1} Q_n \Sigma^{(n)}(z) \prod_{j=1}^{n} [1 + \mathscr{P}_j G^{(j)}(z) [\mathscr{P}_j \Sigma^{(j)}(z)] \mathscr{P} R(z) \mathscr{P}, \tag{5.42}$$

where the associated level-shift operator is

$$\Sigma^{(j)}(z) = V + V Q_j (z - Q_j H Q_j)^{-1} Q_j V, \tag{5.43}$$

and

$$\mathscr{P}_j G^{(j)}(z) \mathscr{P}_j = \mathscr{P}_j \{z - \mathscr{P}_j H_0 - \mathscr{P}_j \Sigma^{(j)}(z) \mathscr{P}_j\}^{-1}. \tag{5.44}$$

Arecchi et al. [23] have applied (5.42) to a system of two two-level atoms. The Hamiltonian for such a system is given by (2.12). We assume that the system was initially in its excited state, i.e.

$$|i\rangle = |1\rangle_1 |1\rangle_2 |\{0\}\rangle. \tag{5.45}$$

The intermediate and the final states are

$$|m_1\rangle = 2^{-1/2}(|1, 2\rangle + |2, 1\rangle)|\{k\}\rangle, \quad |m_2\rangle = 2^{-1/2}(|1, 2\rangle - |2, 1\rangle)|\{k\}\rangle,$$
$$|f\rangle = |2, 2, \{k\,l\}\rangle. \tag{5.46}$$

The projection operators for the present case are

$$\mathscr{P} = |1, 1, \{0\}\rangle \langle 1, 1, \{0\}|, \quad \mathscr{P}_m = |m\rangle \langle m|, \quad m = 1, 2. \tag{5.47}$$

It should be noted that the interaction Hamiltonian allows only one-step transitions. We also assume that the states $|m_1\rangle$ and $|m_2\rangle$ do not interact via the ground state and work in rotating-wave approximation. Then, after calculating the function $G^{(j)}(z)$ in the lowest order of interaction, going over to the continuum limit of field modes, and ignoring the retardation effects, one finds that the probability that there is a photon in mode k and another in l is given by (with r standing for the distance between the two atoms)

$$p_{kl}(\infty) = |g_k g_l|^2 \left\{ \sum_{j=1,2} (1 + (-1)^j \cos \mathbf{k} \cdot \mathbf{r})(1 + (-1)^j \cos \mathbf{l} \cdot \mathbf{r}) \right.$$
$$\cdot \{(\omega_k + \omega_l - 2\omega)^2 + 4\gamma_j^2\} \{(\omega_k - \omega)^2 + \gamma_j^2\}^{-1}$$
$$\cdot \{(\omega_l - \omega)^2 + \gamma_j^2\}^{-1} \{(\omega_k + \omega_l - 2\omega)^2 + 4\gamma^2\}^{-1} \tag{5.48}$$
$$+ 2 \sin(\mathbf{k} \cdot \mathbf{r}) \sin(\mathbf{l} \cdot \mathbf{r}) \, \text{Re} \prod_{j=1,2} (\omega_k + \omega_l - 2\omega + 2i(-1)^j \gamma_j)$$
$$\left. \cdot (\omega_k - \omega + 2i(-1)^j \gamma_j)^{-1} (\omega_l - \omega + 2i(-1)^j \gamma_j)^{-1} \{(\omega_k + \omega_l - 2\omega)^2 + 4\gamma^2\}^{-1} \right\},$$

where

$$\gamma_j = \gamma\{1 + (-1)^j [j_0(\omega r/c) - \tfrac{1}{2} j_2(\omega r/c)]\}, \; \boldsymbol{d} \cdot \boldsymbol{r} = 0$$

and γ is the single-atom decay constant. The extension of (5.48) to the case of N atoms presents difficulties because now one has to construct a complete set of intermediate states and care has to be exercised if transitions between such intermediate states are allowed. The above method, however, has the advantage that it leads directly to the long-time behavior without the need to solve the time-dependent problem.

Finally, we mention that the analysis leading to (5.24) and (5.26) has been generalized by Lambropoulos [24] within the framework of the resolvent operator appropriate to the density operator [cf. Eq. (6.6)].

6. Quantum Statistical Method: Master Equations

We have seen in the earlier sections that the calculation of the properties of the spontaneously emitted radiation gets more and more involved as the number of atoms or the number of atomic levels increases. All the previous methods rested on the basic idea of the "relevant states" involved in the transitions and in such cases one is essentially restricted to the resonant transitions. It appears rather difficult to take into account the virtual transitions which are absolutely essential for considering the Lamb shift. Moreover, the statistical aspects of the emission are obscure — at least, they do not appear to have been discussed in the framework of those theories. In this and subsequent sections we discuss master-equation methods [25, 26], which allow us to study a large class of phenomena involving the interaction of radiation with matter[4, 5]. Our particular emphasis will be on the quantum statistical aspects of the problem. At various stages of this development we will outline the connection with the results of the previous sections and also present a number of new results.

Let ϱ_{A+R} be the density operator characterizing the statistical state of the combined system of the atoms and the radiation field. We introduce the reduced density operators $\varrho_A(t)$ and $\varrho_R(t)$ corresponding to the atomic system and the radiation field, respectively. The reduced density operators are related to ϱ_{A+R} by

$$\varrho_A(t) = \mathrm{Tr}_R \, \varrho_{A+R}(t), \tag{6.1}$$

$$\varrho_R(t) = \mathrm{Tr}_A \, \varrho_{A+R}(t), \tag{6.2}$$

---

[4] For a general review of master-equation methods in quantum optical and related problems, see [27, 28] and literature references cited therein.

[5] Some of the results have been published before in [29 to 33].

where $\mathrm{Tr}_R(\mathrm{Tr}_A)$ denotes the trace over the radiation field (atomic) variables. The density operator $\varrho_{A+R}(t)$ satisfies the Schrödinger equation

$$\dot\varrho_{A+R} = -i[H, \varrho_{A+R}] \equiv -i\mathscr{L}\,\varrho_{A+R}, \tag{6.3}$$

where $\mathscr{L}$ is the Liouville operator defined by

$$\mathscr{L}\cdots = [H, \ldots]. \tag{6.4}$$

A formal solution of (6.3) is

$$\varrho_{A+R}(t) = \exp\{-i\mathscr{L}t\}\,\varrho_{A+R}(0) \equiv \mathrm{e}^{-iHt}\,\varrho_{A+R}(0)\,\mathrm{e}^{iHt}. \tag{6.5}$$

The time-evolution operator and the resolvent operators are given by

$$\mathscr{U}(t) = \mathrm{e}^{-i\mathscr{L}t}, \qquad R(z) = (z - \mathscr{L})^{-1}, \tag{6.6}$$

which replace (5.19) and (5.20).

We are discussing the case of spontaneous emission, hence the initial state of the field is given by

$$\varrho_R(0) = |\{0\}\rangle\langle\{0\}|, \tag{6.7}$$

where $|\{0\}\rangle$ denotes the vacuum of the field. The initial state corresponds to a nonequilibrium situation

$$\varrho_{A+R}(0) = \varrho_A(0)\,\varrho_R(0), \tag{6.8}$$

where $\varrho_A(0)$ is the initial state of the atomic system, which we leave quite arbitrary, since it depends on how the atomic system is prepared. We will obtain an equation of motion for the reduced density operator of the atomic system. It is not possible to obtain a simple master equation for $\varrho_R(t)$ for the reasons given below [following Eq. (7.1)]. To obtain an equation for $\varrho_A(t)$, we will use Zwanzig's projection operator techniques.

We introduce a projection operator $\mathscr{P}$ (which is time-independent) defined by

$$\mathscr{P}\cdots = G\,\mathrm{Tr}_R\cdots. \tag{6.9}$$

The operator G should be such that $\mathscr{P}^2 = \mathscr{P}$ so that

$$\mathrm{Tr}_R\,G = 1. \tag{6.10}$$

The form of G would be dictated by the physical problem and the initial condition. It is clear that

$$\mathscr{P}\varrho_{A+R}(t) = G\varrho_A(t), \tag{6.11}$$

and at time $t = 0$

$$\mathscr{P}\varrho_{A+R}(0) = G\varrho_A(0), \tag{6.12}$$

$$(I - \mathscr{P})\varrho_{A+R}(0) = (\varrho_R(0) - G)\varrho_A(0), \tag{6.13}$$

and if we choose $G = \varrho_R(0)$, then

$$\mathscr{P}\varrho_{A+R}(t) = \varrho_R(0)\varrho_A(t), \quad (1 - \mathscr{P})\varrho_{A+R}(0) = 0, \tag{6.14}$$

so that

$$\varrho_A(t) = \operatorname{Tr}_R \mathscr{P}\varrho_{A+R}(t), \varrho_R(t) = \varrho_R(0) + \operatorname{Tr}_A(1 - \mathscr{P})\varrho_{A+R}(t). \tag{6.15}$$

This choice of G leads to the correct perturbative results (§ 9).

The density operator is now written as

$$\varrho_{A+R}(t) = \mathscr{P}\varrho_{A+R}(t) + (1 - \mathscr{P})\varrho_{A+R}(t), \tag{6.16}$$

where one usually refers to $\mathscr{P}\varrho_{A+R}(t)$ as the relevant operator of the density operator. In the case of spontaneous emission $(1 - \mathscr{P})\varrho_{A+R}$ is as relevant as $\mathscr{P}\varrho_{A+R}(t)$, because the former gives the properties of the emitted radiation. For completeness, we outline here the derivation of the equation for $\mathscr{P}\varrho_{A+R}(t)$. On taking the Laplace transform of (6.5), defined by (5.3), we obtain

$$z\hat{\varrho}_{A+R}(z) - \varrho_{A+R}(0) = -i\mathscr{L}\hat{\varrho}_{A+R}(z). \tag{6.17}$$

On multiplying by $\mathscr{P}$ and $(1 - \mathscr{P})$ we obtain the equations

$$z\mathscr{P}\hat{\varrho}_{A+R} - \mathscr{P}\varrho_{A+R}(0) = -i\mathscr{P}\mathscr{L}\mathscr{P}\hat{\varrho}_{A+R} - i\mathscr{P}\mathscr{L}(1 - \mathscr{P})\hat{\varrho}_{A+R}, \tag{6.18a}$$

$$z(1 - \mathscr{P})\hat{\varrho}_{A+R} - (1 - \mathscr{P})\varrho_{A+R}(0) = -i(1 - \mathscr{P})\mathscr{L}\mathscr{P}\hat{\varrho}_{A+R}$$
$$-i(1 - \mathscr{P})\mathscr{L}(1 - \mathscr{P})\hat{\varrho}_{A+R}. \tag{6.18b}$$

From (6.18b) we have

$$(1 - \mathscr{P})\hat{\varrho}_{A+R} = [z + i(1 - \mathscr{P})\mathscr{L}(1 - \mathscr{P})]^{-1}$$
$$\cdot [(1 - \mathscr{P})\varrho_{A+R}(0) - i(1 - \mathscr{P})\mathscr{L}\mathscr{P}\hat{\varrho}_{A+R}], \tag{6.18c}$$

and then from (6.18a) it follows that

$$z\mathscr{P}\hat{\varrho}_{A+R} - \mathscr{P}\varrho_{A+R}(0) = -i\mathscr{P}\mathscr{L}\mathscr{P}\hat{\varrho}_{A+R} - i\mathscr{P}\mathscr{L}(1 - \mathscr{P})$$
$$\cdot [z + i(1 - \mathscr{P})\mathscr{L}(1 - \mathscr{P})]^{-1}[(1 - \mathscr{P})\varrho_{A+R}(0) - i(1 - \mathscr{P})\mathscr{L}\mathscr{P}\hat{\varrho}_{A+R}]. \tag{6.18d}$$

On taking the inverse Laplace transform of (6.18c) and (6.18) we obtain the equations

$$(\partial/\partial t)\,\mathscr{P}\varrho_{A+R}(t) + i\mathscr{P}\mathscr{L}\mathscr{P}\varrho_{A+R}(t)$$

$$= -i\mathscr{P}\mathscr{L}(1-\mathscr{P})\exp\{-i(1-\mathscr{P})\mathscr{L}(1-\mathscr{P})t\}(1-\mathscr{P})\varrho_{A+R}(0) \qquad (6.19)$$

$$- \mathscr{P}\mathscr{L}(1-\mathscr{P})\int_0^t d\tau\exp\{-i(1-\mathscr{P})\mathscr{L}(1-\mathscr{P})\tau\}(1-\mathscr{P})\mathscr{L}\mathscr{P}\varrho_{A+R}(t-\tau),$$

$$(1-\mathscr{P})\varrho_{A+R}(t) = \exp\{-i(1-\mathscr{P})\mathscr{L}(1-\mathscr{P})t\}(1-\mathscr{P})\varrho_{A+R}(0) \qquad (6.20)$$

$$- i\int_0^t d\tau\exp\{-i(1-\mathscr{P})\mathscr{L}(1-\mathscr{P})\tau\}(1-\mathscr{P})\mathscr{L}\mathscr{P}\varrho_{A+R}(t-\tau).$$

Because of our special choice of projection operator, these equations simplify to

$$(\partial/\partial t)\,\mathscr{P}\varrho_{A+R}(t) + i\mathscr{P}\mathscr{L}\mathscr{P}\varrho_{A+R}(t) + \mathscr{P}\mathscr{L}(1-\mathscr{P})\int_0^t d\tau \qquad (6.21)$$

$$\cdot\exp[-i(1-\mathscr{P})\mathscr{L}(1-\mathscr{P})\tau](1-\mathscr{P})\mathscr{L}\mathscr{P}\varrho_{A+R}(t-\tau) = 0,$$

$$(1-\mathscr{P})\varrho_{A+R}(t) = -i\int_0^t d\tau\exp[-i(1-\mathscr{P})\mathscr{L}(1-\mathscr{P})\tau] \qquad (6.22)$$

$$\cdot(1-\mathscr{P})\mathscr{L}\mathscr{P}\varrho_{A+R}(t-\tau).$$

Once $\mathscr{P}\varrho_{A+R}(t)$ is known from the solution of (6.21), $(1-\mathscr{P})\varrho_{A+R}(t)$ can be obtained from (6.22). We will now write the interaction Hamiltonian and the corresponding Liouville operator as

$$H = H_A + H_R + H_{AR}, \qquad \mathscr{L} = \mathscr{L}_A + \mathscr{L}_R + \mathscr{L}_{AR}.$$

It is evident that

$$\mathscr{P}\mathscr{L}_A = \mathscr{L}_A\mathscr{P}, \qquad \mathscr{P}\mathscr{L}_R = \mathscr{L}_R\mathscr{P} = 0. \qquad (6.23)$$

Moreover, since $\mathscr{L}_{AR}$ is linear in a_{ks} and a_{ks}^+, it follows that

$$\mathscr{P}\mathscr{L}_{AR}\mathscr{P}\cdots = 0. \qquad (6.24)$$

On using (6.23) and (6.24), (6.21) simplifies to

$$(\partial/\partial t)\,\mathscr{P}\varrho_{A+R}(t) + i\mathscr{L}_A\mathscr{P}\varrho_{A+R}(t) + \mathscr{P}\mathscr{L}_{AR}(1-\mathscr{P})\int_0^t d\tau$$

$$\cdot\exp[-i(1-\mathscr{P})\mathscr{L}(1-\mathscr{P})\tau](1-\mathscr{P})\mathscr{L}_{AR}\mathscr{P}\varrho_{A+R}(t-\tau),$$

or

$$(\partial/\partial t)\,\mathscr{P}\varrho_{A+R}(t) + i\mathscr{L}_A\mathscr{P}\varrho_{A+R}(t) + \mathscr{P}\mathscr{L}_{AR}\int_0^t d\tau\,U_0(\tau)$$

$$\cdot(1-\mathscr{P})\,U(\tau)\,\mathscr{L}_{AR}\mathscr{P}\varrho_{A+R}(t-\tau) = 0, \qquad (6.25)$$

where

$$U_0(\tau) = \exp\{-i\tau(\mathscr{L}_A + \mathscr{L}_R)\}, \tag{6.26a}$$

$$U(\tau) = T \exp\left\{-i\int_0^\tau dt'(1-\mathscr{P})\,U_0(-t')\,\mathscr{L}_{AR}U_0(t')\,(1-\mathscr{P})\right\}, \tag{6.26b}$$

and where T is the time-ordering operator. It should be noted that, so far, we have not made any approximation as to the strength of the interaction between the field and matter. The second term in (6.25) is of at least second order in interaction. The lowest order of approximation (Born approximation) is obtained by letting $U(\tau) \to 1$ and then (6.25) reduces to

$$(\partial/\partial t)\,\mathscr{P}\varrho_{A+R}(t) + i\mathscr{L}_A\mathscr{P}\varrho_{A+R}(t)$$
$$+ \mathscr{P}\mathscr{L}_{AR}\int_0^t d\tau\, U_0(\tau)\,\mathscr{L}_{AR}\mathscr{P}\varrho_{A+R}(t-\tau) = 0. \tag{6.27}$$

This equation is still an integro-differential equation, i.e. the time derivative of $\mathscr{P}\varrho_{A+R}$ at time t depends on the value of $\mathscr{P}\varrho_{A+R}$ at all the earlier times. The Born approximation means that the emitted photon does not react back on the atom. On transforming (6.27) to the interaction picture we obtain

$$(\partial/\partial t)\,\{\mathscr{P}\varrho_{A+R}^I(t)\} + \int_0^t d\tau\,\mathscr{P}\mathscr{L}_{AR}^I(t)\,\mathscr{L}_{AR}^I(t-\tau)\,\mathscr{P}\varrho_{A+R}^I(t-\tau) = 0, \tag{6.28}$$

where the superscript I stands for the operators in the interaction picture.

For the case of a collection of identical two-level atoms we have from (2.12)

$$\mathscr{L}_{AR}^I(t) = \sum_{ksj} g_{ksj}\,[a_{ks}(S_j^+\,e^{i\omega t} + S_j^-\,e^{-i\omega t})\,e^{-i\omega_{ks}t}, \ldots]. \tag{6.29}$$

On substituting (6.29) in (6.28) and on using

$$\text{Tr}\{\varrho_R(0)\,a_{ks}^+\,a_{k's'}\} = 0, \qquad \text{Tr}\{\varrho_R(0)\,a_{ks}\,a_{k's'}^+\} = \delta_{kk'}\delta_{ss'},$$

$$\text{Tr}\{\varrho_R(0)\,a_{ks}\,a_{k's'}\} = \text{Tr}\{\varrho_R(0)\,a_{ks}^+\,a_{k's'}^+\} = 0,$$

we obtain

$$(\partial/\partial t)\,\varrho_A^I(t) + \sum_{ksjl}\left[g_{ksj}\,g_{ksl}^*\int_0^t d\tau\,e^{-i\omega_{ks}\tau}\{e^{i\omega\tau}[S_j^+, S_l^-\,\varrho_A^I(t-\tau)]\right.$$
$$\left. + e^{-i\omega\tau}[S_j^-, S_l^+\,\varrho_A^I(t-\tau)]\} + \text{H.C.}\right] = 0. \tag{6.30}$$

In obtaining (6.30) we also made the rotating-wave approximation (RWA), i.e. we ignored terms like $S_i^+ S_j^+$, $S_i^- S_j^-$. The RWA on (6.30) is very different from the RWA on the Hamiltonian itself, as we discuss more fully in Appendix A.

On taking the Laplace transform, (6.30) reduces to (ignoring the suffices I, A)

$$z\hat{\varrho} - \varrho(0) + \sum_{ij} \hat{\gamma}_{ij}^+ \{S_i^+ S_j^- \hat{\varrho} - 2S_j^- \hat{\varrho} S_i^+ + \hat{\varrho} S_i^+ S_j^-\}$$

$$\tag{6.31}$$

$$+ \sum_{ij} \hat{\gamma}_{ij}^- \{S_i^- S_j^+ \hat{\varrho} - 2S_j^+ \hat{\varrho} S_i^- + \hat{\varrho} S_i^- S_j^+\} + i \sum_{ij} \hat{\Omega}_{ij} [S_i^+ S_j^-, \hat{\varrho}] = 0,$$

where

$$\hat{\gamma}_{ij}^\pm(z) = \sum_{ks} g_{ksi} g_{ksj}^* z \{z^2 + (\omega_{ks} \mp \omega)^2\}^{-1}, \tag{6.32a}$$

$$\hat{\Omega}_{ii}(z) = - \sum_{ks} |g_{ks}|^2 \{(\omega_{ks} - \omega)[z^2 + (\omega_{ks} - \omega)^2]^{-1} - (\omega \to - \omega)\}, \tag{6.32b}$$

$$\hat{\Omega}_{ij}(z) = - \sum_{ks} g_{ksi} g_{ksj}^* \{(\omega_{ks} - \omega)[z^2 + (\omega_{ks} - \omega)^2]^{-1} + (\omega \to - \omega)\}. \tag{6.32c}$$

$$(i \neq j)$$

In obtaining (6.31) we also used the relation $[S_i^+ S_i^-, \varrho] = - [S_i^- S_i^+, \varrho]$. We will now take the limit $L^3 \to \infty$ so that the summation over k should be replaced by an integral over the continuum of modes, i.e.

$$\sum_{ks} \to (L^3/(2\pi)^3) \int d^3 k \sum_s. \tag{6.33}$$

We now make one further approximation, the *Markov* approximation, i.e. we ignore the retardation effects and take the long time limit i.e.

$$t \gg 1/\omega, \quad t \gg \max(r_{ij}/c), \tag{6.34}$$

and if the passage time of the light is small compared to the time Δt in which appreciable changes occur, i.e.

$$\max(r_{ij}/c) \ll \Delta t, \tag{6.35}$$

then one can replace $\hat{\gamma}$ and $\hat{\Omega}$ in (6.31) by their limiting values as $z \to 0^+$. In the present context the Markov approximation can only be made after the limits (6.33) to (6.35) are taken. Hence, under the Markov approximation and with the long time limit, (6.31) reduces to

$$\partial \varrho/\partial t = - i \sum_{ij} \Omega_{ij} [S_i^+ S_j^-, \varrho] - \sum_{ij} \gamma_{ij} \{S_i^+ S_j^- \varrho - 2S_j^- \varrho S_i^+ + \varrho S_i^+ S_j^-\}, \tag{6.36}$$

where[6]

$$\gamma_{ij} = \lim_{z \to 0^+} \hat{\gamma}_{ij}^+(z) = \pi \sum_{ks} g_{ksi} g_{ksj}^* \delta(\omega - \omega_{ks}), \tag{6.37}$$

$$\Omega_{ii} = \lim_{z \to 0^+} \hat{\Omega}_{ii}(z) = - \sum_{ks} |g_{ks}|^2 \{(\omega_{ks} - \omega)^{-1} - (\omega_{ks} + \omega)^{-1}\}, \tag{6.38}$$

$$\Omega_{ij} = \lim_{z \to 0^+} \hat{\Omega}_{ij}(z) = - \sum_{ks}' g_{ksi} g_{ksj}^* \{(\omega_{ks} - \omega)^{-1} + (\omega_{ks} + \omega)^{-1}\}, \quad (i \neq j). \tag{6.39}$$

On transforming (6.36) to the Schrödinger picture, we have

$$\partial \varrho / \partial t = - i \sum_i (\omega + \Omega_{ii}) [S_i^z, \varrho] - i \sum_{i \neq j} \Omega_{ij} [S_i^+ S_j^-, \varrho]$$
$$- \sum_{ij} \gamma_{ij} (S_i^+ S_j^- \varrho - 2 S_j^- \varrho S_i^+ + \varrho S_i^+ S_j^-). \tag{6.40}$$

In deriving (6.40) we took into account only the $-d \cdot E$ interaction. We must now take into account the H_{self} as given by (2.20). On adding the effect of H_{self}, (6.40) reduces to

$$\partial \varrho / \partial t = - i \sum_i (\omega + \Omega_{ii}) [S_i^z, \varrho] - i \sum_{i \neq j} (\Omega_{ij} + V_{ij} + \mathcal{V}_{ij}) [S_i^+ S_j^-, \varrho]$$
$$- \sum_{ij} \gamma_{ij} (S_i^+ S_j^- \varrho - 2 S_j^- \varrho S_i^+ + \varrho S_i^+ S_j^-). \tag{6.41}$$

We will now examine the values of the coefficients which appear in (6.41). We first note that γ_{ij} is given by

$$\gamma_{ij} = 2\pi^2 c(2\pi)^{-3} \int k^3 \, dk \, d\Omega' \delta(\omega - kc) |d|^2 (1 - \cos^2 \theta') e^{i\mathbf{k} \cdot \mathbf{r}_{ij}}, \tag{6.42}$$

where as before

$$r_{ij} = r_i - r_j. \tag{6.43}$$

The angular part which we need in many calculations is obtained by using (2.22)

$$I \equiv \int d\Omega' \sin^2 \theta' e^{i\mathbf{k} \cdot \mathbf{R}} = 4\pi \{ \tfrac{2}{3} j_0(kR) + (\cos^2 \theta - \tfrac{1}{3}) j_2(kR) \}, \tag{6.44}$$

where θ is the angle between $\mathbf{d}$ and $\mathbf{R}$. On substituting (6.44) in (6.42) we obtain for γ_{ij}:

$$\gamma_{ij} = \gamma \{ j_0(k_0 r_{ij}) + (\tfrac{3}{2} \cos^2 \theta - \tfrac{1}{2}) j_2(k_0 r_{ij}) \}, \tag{6.45}$$

where

$$\gamma = \tfrac{2}{3} |d|^2 \omega^3 / c^3, \quad k_0 = \omega / c. \tag{6.46}$$

[6] Throughout this article denominators like $(k \mp k_0)^{-1}$ should be interpreted in the principal value sense.

The evaluation of Ω_{ii} is an involved problem. It is connected with the Lamb shift and is given by

$$\Omega_{ii} = \tfrac{2}{3}|d|^2\pi^{-1}\int k^3\,dk\{(k+k_0)^{-1} - (k-k_0)^{-1}\} \tag{6.47}$$
$$= -\tfrac{4}{3}|d|^2\pi^{-1}k_0\int k\,dk - \tfrac{2}{3}|d|^2\pi^{-1}k_0^3\int dk\{(k+k_0)^{-1} + (k-k_0)^{-1}\}\,.$$

The first term is quadratically divergent and the second term has logarithmic divergence. The quadratic divergence can be removed by using the free-electron Hamiltonian [15]. The remaining term can be written as

$$\Omega_{ii} = -(\gamma/\pi)\ln\{|\omega_c/\omega - 1|\,(\omega_c/\omega + 1)\}\,, \tag{6.48}$$

where ω_c is the cutoff frequency.

We now turn to the calculation[7] of Ω_{ij}:

$$\Omega_{ij} = -\tfrac{2}{3}|d|^2\pi^{-1}\int k^3\,dk[(k-k_0)^{-1} + (k+k_0)^{-1}]$$
$$\cdot[j_0(kr_{ij}) + (\tfrac{3}{2}\cos^2\theta - \tfrac{1}{2})j_2(kr_{ij})]$$
$$= -\tfrac{4}{3}|d|^2\pi^{-1}\int k^2\,dk[j_0(kr_{ij}) + (\tfrac{3}{2}\cos^2\theta - \tfrac{1}{2})j_2(kr_{ij})] \tag{6.49}$$
$$- \tfrac{2}{3}|d|^2\pi^{-1}k_0\int k^2\,dk\{(k-k_0)^{-1} - (k+k_0)^{-1}\}$$
$$\cdot[j_0(kr_{ij}) + (\tfrac{3}{2}\cos^2\theta - \tfrac{1}{2})j_2(kr_{ij})]\,,$$

Note that the first integral in (6.49) is identical to $\mathscr{V}_{ij}$ [cf. Eq. (2.23)] and hence

$$\Omega_{ij} + \mathscr{V}_{ij} = -\tfrac{2}{3}|d|^2\pi^{-1}k_0^2\int k\,dk\{(k-k_0)^{-1} + (k+k_0)^{-1}\}$$
$$\cdot[j_0(kr_{ij}) + (\tfrac{3}{2}\cos^2\theta - \tfrac{1}{2})j_2(kr_{ij})]\,. \tag{6.50}$$

It should be noticed that the integrand in (6.50) has poles at $k = \pm k_0, 0$. The singularity at $k = 0$ comes from $j_2(kr_{ij})$. It can be shown by contour integration that the contribution from $k = 0$ is precisely $-V_{ij}$ and thus cancels the dipole–dipole interaction term and hence[8]

$$\Omega_{ij} + \mathscr{V}_{ij} + V_{ij} = \gamma\Delta_{ij}(k_0 r_{ij})\,, \tag{6.51}$$

$$\Delta_{ij}(x) = \tfrac{3}{2}\{(1 - 3\cos^2\theta)\,[(\sin x)/x^2 + (\cos x)/x^3]$$
$$- (1 - \cos^2\theta)\,(\cos x)/x\}\,. \tag{6.52}$$

---

[7] The relevant k integrations can be done using

$$\mathrm{P}\int_{-\infty}^{+\infty}\frac{\cos kR\,dk}{(k^2 - k_0^2)} = -(\pi/k_0)\sin k_0 R, \qquad \mathrm{P}\int_{-\infty}^{+\infty}\frac{\sin kR\,dk}{(k^2 - k_0^2)} = 0,$$

with P standing for the principal part.

[8] The term $\mathscr{V}_{ij} + V_{ij}$ makes no effective contribution, as it is proportional to $\delta(r_i - r_j)\,(i \neq j)$. We have followed this complicated route to facilitate comparison with the case when $-A \cdot p$ interaction is used.

On substituting (6.51) and (6.45) in (6.41), the master equation reduces to

$$\partial \varrho / \partial t = -i\omega_0 \sum_i [S_i^z, \varrho] - i \sum_{i \neq j} \Omega_{ij} [S_i^+ S_j^-, \varrho]$$

$$- \sum_{ij} \gamma_{ij} (S_i^+ S_j^- \varrho - 2 S_j^- \varrho S_i^+ + \varrho S_i^+ S_j^-),$$

$\qquad\qquad\qquad\qquad\qquad\qquad\qquad\qquad\qquad\qquad\qquad$ (6.53)

where ω_0 is the renormalized frequency which is equal to the sum of the old frequency and Ω_{ii} (in most of the formulae we will ignore the subscript from ω_0), γ_{ij} is given by (6.45), and Ω_{ij} is now given by (6.52). This is the final form of the master equation[9] and it will play a basic role in our further development.

It is interesting to note that the master equation is the same, whether we work with the interaction (2.12) or (2.15), as long as we make the rotating-wave approximation and the Markov approximation. A straightforward analysis shows that the master equation (6.53) is still obtained with

$$\gamma_{ij} = (\pi \omega^2 / c^2) \sum g_{ksi} g_{ksj}^* \delta(\omega - \omega_{ks}) (1/k^2), \qquad\qquad (6.54\,\text{a})$$

$$\Omega_{ii} = -(\omega^2 / c^2) \sum |g_{ks}|^2 (1/k^2) [(\omega_{ks} - \omega)^{-1} - (\omega_{ks} + \omega)^{-1}], \qquad (6.54\,\text{b})$$

$$\Omega_{ij} = -(\omega^2 / c^2) \sum g_{ksi} g_{ksj}^* (1/k^2) [(\omega_{ks} - \omega)^{-1} + (\omega_{ks} + \omega)^{-1}] + V_{ij}. \,(6.54\,\text{c})$$

The explicit form of γ_{ij} would be the same as (6.45) since in the integration, the delta function picks up the value only on the energy shell. On com-

---

[9a] The master-equation (6.53) has been derived by Lehmberg [34] using direct integration of the Heisenberg equations of motion. We discuss this method in § 8. It has also been obtained in [35].

[9b] This master equation was originally obtained by the author [29] using Schwinger's boson representation [36] in conjunction with phase-space methods [37]. The derivation of [29] is an interesting illustration of Schwinger's boson representation in a dynamical context.

[9c] The retardation effects can also be included in the present master equation. For other treatments see [38].

[9d] The density operator is a semipositive-definite operator. It remains to be proved that the solution of the master equation is a semipositive-definite operator. This is an involved question. In the case of a Markovian master equation (6.53) it is shown in [35] that the positive definiteness is preserved if the matrix γ is a semipositive-definite matrix. For a single atom the non-Markovian master equation (6.31) leads to a nonpositive-definite density operator (cf. [39]).

[9e] Picard and Willis [40] have obtained extra terms in the master equation. They claim that we assume that the radiation field remains in the vacuum state all the time, which is contrary to what we show in § 7, 10, and 15. The misunderstanding seems to be due to the form of the projection operator (6.12) we used in deriving the master equation. All we need to assume is that the radiation field is in the vacuum state at time $t = 0$ (random phase condition at $t = 0$, in the old language).

parison we see that $\Omega_{ij} + \mathscr{V}_{ij} + V_{ij}$ as given by (6.50) is the same as Ω_{ij} given by (6.54c). The term Ω_{ii} needs some attention:

$$\begin{aligned}\Omega_{ii} &= \tfrac{2}{3}|d|^2\pi^{-1}k_0^2 \int k\,\mathrm{d}k[(k+k_0)^{-1} - (k-k_0)^{-1}] \\ &= -\tfrac{2}{3}|d|^2\pi^{-1}k_0^3 \int \mathrm{d}k[(k+k_0)^{-1} + (k-k_0)^{-1}]\,,\end{aligned} \tag{6.54d}$$

which is identical to the second term in (6.47). Thus, with the formalism using the $-A \cdot p$ interaction, there is no quadratically divergent term. Power and Zienau have already remarked that the $-d \cdot E$ interaction is not well suited for the calculation of the Lamb shift[10]. Although the master equation is independent of whether $-A \cdot p$ or $-d \cdot E$ is used, we will see later that the difference in these two interactions results in different expressions for the line shape [cf. (3.13), (3.14)].

For the case when the atoms were confined to a region smaller than a wavelength $k_0 R \ll 1$, on using the expansions

$$j_0(x) = 1 + O(x^2)\,, \quad j_2(x) = O(x^2)\,, \tag{6.55}$$

we find

$$\gamma_{ij} \approx \gamma + O(r_{ij}^2)\,, \quad \Omega_{ij} \approx V_{ij}\,, \tag{6.56}$$

which is as expected: the retarded dipole–dipole interaction Ω_{ij} is replaced by the static dipole-dipole interaction.

The coefficients Ω_{ij} and γ_{ij} that appear in the master equation are related to the shifts and the widths of the levels. These coefficients have the value one would obtain by a second-order perturbation theory involving the exchange of a photon (both real and virtual). These parameters are also closely related to the real and imaginary parts of the *free-field propagator* (see e.g. [42])

$$\begin{aligned}\mathscr{D}_E &\equiv [d \cdot E(x, t), d \cdot E(x', t')] = (2\pi c/L^3)\,|d|^2 \sum k^3\,\mathrm{d}k\,\mathrm{d}\Omega' \\ &\quad \cdot (1 - \cos^2\theta')\,\mathrm{e}^{ik \cdot (x-x')}\{\mathrm{e}^{-ikc(t-t')} - \mathrm{e}^{ikc(t-t')}\} \tag{6.57} \\ &\equiv \mathscr{D}_E(x - x', t - t')\,.\end{aligned}$$

The one-sided, Fourier transform of $\mathscr{D}_E$ is given by

$$\begin{aligned}\mathscr{D}_E(R, \omega) &= \int\limits_0^\infty \mathrm{d}t\,\mathrm{e}^{i\omega t}\,\mathscr{D}_E(R, t) \\ &= (|d|^2 c/4\pi^2) \int k^3\,\mathrm{d}k\{\delta_-(\omega_k - \omega) - \delta_+(\omega_k + \omega)\} \\ &\quad \cdot \int \mathrm{e}^{ik \cdot R}\,\mathrm{d}\Omega'(1 - \cos^2\theta') = (|d|^2 c/4\pi)\int k^3\,\mathrm{d}k\delta(\omega_k - \omega) \tag{6.58} \\ &\quad \cdot \int \mathrm{e}^{ik \cdot R}\,\mathrm{d}\Omega'(1 - \cos^2\theta') - (i|d|^2 c/4\pi^2)\int k^3\,\mathrm{d}k \\ &\quad \cdot \{(\omega_k - \omega)^{-1} + (\omega_k + \omega)^{-1}\}\int \mathrm{e}^{ik \cdot R} \cdot \sin^2\theta'\,\mathrm{d}\Omega' = \gamma_{ij} + i\Omega_{ij}\,,\end{aligned}$$

---

[10] The fact that Ω_{ii} has a well-defined value if a $-A \cdot p$ interaction is used (i.e. needs no renormalization) for the case of a single two-level atom has also been discussed in [41].

where γ_{ij} and Ω_{ij} are given by (6.37) and (6.39), respectively. Stephen has also referred to Ω_{ij} as the first-order dispersion force between two atoms [43].

The master equation (6.53) contains complete information about the atomic system. It can for example, be used to discuss how the atomic system decays, or to solve problems concerning population fluctuations, development of atomic correlations, etc. The solution of the master equation immediately gives the one-time expectation values. To obtain two-time or multitime correlation functions, we will use the quantum regression theorem, assuming that the behavior of the system is Markovian. The regression theorem states that if one-time mean values are expressed as

$$\langle Q_i(t_1)\rangle = \sum f_{i\alpha}(t_1 - t_2)\langle Q_\alpha(t_2)\rangle, \qquad (t_1 > t_2), \tag{6.59}$$

where $f_{i\alpha}$ are the numerical functions and $Q_\alpha(t_2)$ form a complete set of operators, then the multitime correlations are given by

$$\langle Q_j(t_1')Q_i(t_1)Q_l(t_1'')\rangle = \sum f_{i\alpha}(t_1 - t_1')\langle Q_j(t_1')Q_\alpha(t_1')Q_l(t_1'')\rangle, \qquad t_1'' < t' < t. \tag{6.60}$$

In particular, for the two-time correlations we have

$$\langle Q_j(t_1')Q_i(t_1)\rangle = \sum f_{i\alpha}(t_1 - t_1')\langle Q_j(t_1')Q_\alpha(t_1')\rangle. \tag{6.61}$$

The one-time mean values appearing in (6.59) or (6.61) are, of course, to be obtained from the solution of the master equation. For the proof of this important theorem, see refs. [44 to 46].

From (6.53) we find that the mean value of any atomic operator obeys the equation

$$\partial\langle Q\rangle/\partial t = i\omega_0 \sum_i \langle [S_i^z, Q]\rangle + i\sum_{i\neq j}\Omega_{ij}\langle [S_i^+ S_j^-, Q]\rangle$$
$$- \sum_{ij}\gamma_{ij}\{\langle [Q, S_i^+]S_j^-\rangle + \langle S_i^+[S_j^-, Q]\rangle\}. \tag{6.62}$$

In particular, the dipole moment of the ith atom obeys the equations

$$(\partial/\partial t)\langle S_i^+\rangle = (i\omega_0 - \gamma)\langle S_i^+\rangle + 2\sum_{i\neq j}(\gamma_{ij} - i\Omega_{ij})\langle S_i^z S_j^+\rangle, \tag{6.63}$$

$$(\partial/\partial t)\langle S_i^z\rangle = -\sum_j\gamma_{ij}(\langle S_i^+ S_j^-\rangle + \langle S_j^+ S_i^-\rangle) - i\sum_{j\neq i}\Omega_{ij}(\langle S_i^+ S_j^-\rangle - \text{c.c.}), \tag{6.64}$$

where we made use of the relations (2.11).

Finally, we mention that if each two-level atom is replaced by a harmonic oscillator [the interaction Hamiltonian (2.25)], then the master equation (6.53) should be replaced by

$$\partial \varrho / \partial t = -i\omega_0 \sum [a_i^+ a_i, \varrho] - i \sum_{i \neq j} \Omega_{ij} [a_i^+ a_j, \varrho]$$
$$- \sum_{ij} \gamma_{ij} (a_i^+ a_j \varrho - 2a_j \varrho a_i^+ + \varrho a_i^+ a_j), \tag{6.65}$$

where Ω_{ij}, γ_{ij} have the same value as before. The renormalized frequency for this model is given by

$$\omega_0 = \omega + \Omega_{ii}^{(-)}$$
$$\Omega_{ii}^{(-)} = -\tfrac{2}{3}(|d|^2/\pi) \int k^3 \, dk \, [(k+k_0)^{-1} + (k-k_0)^{-1}]$$
$$= (\gamma/\pi) \{\ln [(\omega_c/\omega + 1)/(\omega_c/\omega - 1)] - 2\omega_c/\omega\} \approx -2\gamma\omega_c/\pi\omega, \tag{6.66}$$

which one also finds by using the Bethe formula [47] specialized to the case of a harmonic oscillator. The term $-\tfrac{4}{3}(|d|^2/\pi) \int k^2 \, dk$ cancels with the contribution from the contact interaction $2\pi \int |P^\perp|^2 \, d^3r$ if we make the RWA. Note also that

$$\Omega_{ii}^{(-)} = \lim_{r_{ij} \to 0} \Omega_{ij} = \text{Im} \int_0^\infty e^{i\omega t} \, \mathscr{D}_E(0, t) \, dt. \tag{6.67}$$

The Lamb shift for the two-level model cannot be obtained so simply from the free-field propagator, which is connected with the spin $-\tfrac{1}{2}$ nature of the dipole-moment operator. It has been suggested by Bullough and Caudery [48] that in place of $\mathscr{D}_E$ [defined by (6.51)], which is a number, one should work with

$$Q_E \equiv \langle \{0\} | \{d \cdot E(x, t), d \cdot E(x', t')\}_+ | \{0\} \rangle, \tag{6.68}$$

i.e. with the expectation value of the anticommutator of the field operators at the points. On evaluating (6.68) we obtain

$$Q_E(R, t) = 2\pi c (2\pi)^{-3} |d|^2 \int k^3 \, dk \int e^{ik \cdot R} \sin^2 \theta' \, d\Omega' \cdot \{e^{-ikct} + e^{ikct}\}. \tag{6.69}$$

The one-sided Fourier transform of Q_E is

$$Q_E(R, \omega) = (|d|^2 c/4\pi) \int k^3 \, dk \, \delta(\omega_k - \omega) \int e^{ik \cdot R} d\Omega' \sin^2 \theta'$$
$$- (i |d|^2 c/4\pi^2) \int k^3 \, dk \, \{(\omega_k - \omega)^{-1} - (\omega_k + \omega)^{-1}\} \int e^{ik \cdot R} d\Omega' \sin^2 \theta',$$

and hence

$$Q_E(0, \omega) = \gamma + i\Omega_{ii}, \tag{6.70}$$

where γ and Ω_{ii} are given by (6.46) and (6.47), respectively.

For $R \neq 0$, $Q_E(R, \omega)$ is rather complicated

$$Q_E(R, \omega) = \gamma_{ij} + (2i/\pi) \, \boldsymbol{d} \cdot \boldsymbol{V} \times \boldsymbol{V} \times \boldsymbol{d} \, \psi(k_0 R)/R \,, \tag{6.71a}$$

where

$$\psi(x) = \sin(x) \, \mathrm{Ci}(x) - \cos(x) \, \mathrm{Si}(x) \,, \tag{6.71b}$$

and where Ci and Si are the standard integrals

$$\mathrm{Ci}(x) = - \int_x^\infty \cos x \, \mathrm{d}x/x \,, \quad \mathrm{Si}(x) = \int_0^x \sin x \, \mathrm{d}x/x \,.$$

It is perhaps of some interest to note that in the Markovian master equation only the real part of Q_E appears for $R \neq 0$. To elucidate the role played by the expectation values of the commutator and the anticommutator of the field operator at two different space–time points, we consider the master Eq. (6.28). On substituting for $\mathscr{L}_{AR}^I(t)$ the expression

$$\mathscr{L}_{AR}^I(t) = - \sum_i [\boldsymbol{d} \cdot \boldsymbol{E}(\boldsymbol{r}_i, t) \, p_i(t)] \,, \tag{6.72}$$

where

$$p_i(t) = S_i^+ \, e^{i\omega t} + S_i^- \, e^{-i\omega t} \,, \tag{6.73}$$

and on simplification, we find that (6.28) can be written as

$$\partial \varrho_A^I/\partial t + \int_0^t \mathrm{d}\tau \, K(t, \tau) = 0 \,, \tag{6.74}$$

where

$$\begin{aligned}
K(t, \tau) = \sum_{ij} \tfrac{1}{2} Q_E(\boldsymbol{r}_{ij}, \tau) \, [p_i(t), [p_j(t - \tau), \varrho_A^I(t - \tau)]] \\
+ \tfrac{1}{2} \mathscr{D}_E(\boldsymbol{r}_{ij}, \tau) \, [p_i(t), \{p_j(t - \tau), \varrho_A^I(t - \tau)\}_+] \,.
\end{aligned} \tag{6.75}$$

In (6.75) Q_E and $\mathscr{D}_E$ are given by (6.68) and (6.57), respectively. It is clear from the above that both commutator and anticommutator figure in the master equation. In the special cases when one is interested in the equations of motion for certain specific variables, then only one of the two would contribute. For example, for the case of a single atom, it is easily found on using (6.75) that $\langle S_i^- \rangle$ obeys the equation

$$(\partial/\partial t) \langle S_i^- \rangle + i\omega \langle S_i^- \rangle + \int_0^t Q_E(\tau) \langle S_i^-(t - \tau) \rangle \, \mathrm{d}\tau = 0 \,, \tag{6.76}$$

and thus only Q_E makes a contribution. On the other hand, for the oscillator we find

$$(\partial/\partial t) \langle a \rangle + i\omega \langle a \rangle + \int_0^t \mathscr{D}_E(\tau) \langle a(t - \tau) \rangle \, \mathrm{d}\tau = 0 \,. \tag{6.77}$$

In deriving (6.76) and (6.77) we also made RWA. Equations (6.76) and (6.77) clearly show how the memory kernel switches from Q_E to $\mathscr{D}_E$ when the atom is replaced by a harmonic oscillator. The memory kernel for an arbitrary function of atomic operators or oscillator operators involves both Q_E and $\mathscr{D}_E$, as is evident from (6.74). The Eqs. (6.76) and (6.77) coincide with those of Bullough and Caudery, who developed a kind of perturbation theory in which they picked up the terms involving either anticommutators or commutators, depending on the variable under consideration.

We also comment on the retardation effects which we ignored in deriving (6.41). For the sake of simplicity, we consider only the case of two two-level atoms for a time interval such that $t < r/c$ (r being the distance between two atoms). For such times $\mathscr{D}_E$ vanishes and only the term Q_E makes a contribution to $K(t, \tau)$. Let ϱ_i^I be the reduced-density operator corresponding to the ith atom. This is obtained by taking the trace over the other atom. The double commutator $[p_1(t), [p_2(t - \tau), \varrho_A^I(t - \tau)]]$ will not contribute if the trace is taken over the other atom. Hence, for $t < r/c$ the reduced-density operator obeys the equation

$$\partial \varrho_i^I / \partial t + \int_0^t d\tau \{\tfrac{1}{2} Q_E(0, \tau) [p_i(t), [p_i(t - \tau), \varrho_i^I(t - \tau)]]$$
$$+ \tfrac{1}{2} \mathscr{D}_E(0, \tau) [p_i(t), \{p_i(t - \tau), \varrho_i^I(t - \tau)\}]\} = 0, \tag{6.78}$$

showing that the reduced-density operator for each atom is the same as if the second atom were absent.

Finally, for the harmonic oscillator model the equation of motion for the dipole operator is extremely simple, viz.

$$\langle \dot{a}_i \rangle = - i\omega \langle a_i \rangle - i \sum_{j \neq i} \Omega_{ij} \langle a_j \rangle - \sum_j \gamma_{ij} \langle a_j \rangle, \tag{6.79}$$

i.e. the equations are linear.

7. Quantum Statistical Properties of the Radiation Field

In the previous section we obtained the master equation for the reduced-density operator of the atomic system. The natural question now is: What are the statistical properties of the radiation which is emitted from a collection of atoms or molecules? We have already presented in Chapter 5 the relation between the field-density operator and the atomic-density operator. On combining (6.15) and (6.22) we obtain

$$\varrho_R(t) = \varrho_R(0) - i \operatorname{Tr}_A \int_0^t d\tau \exp\{- i(1 - \mathscr{P}) \mathscr{L} (1 - \mathscr{P}) \tau\} \mathscr{L} \varrho_R(0) \varrho_A(t - \tau). \tag{7.1}$$

This in principle enables us to determine $\varrho_R(t)$ from the knowledge of $\varrho_A(\tau)$. In practice it appears difficult to use (7.1) because of the presence of the factor $\exp\{-i(1-\mathscr{P})\mathscr{L}(1-\mathscr{P})\tau\}$ under the integral and also partly because of the fact that a Markov approximation cannot be made for the field density operator and even the validity of the Born approximation on (7.1) appears doubtful. It was for these reasons that we obtained the master equation for the atomic system. To determine the properties of the field, we must follow a different route.

We start from the Heisenberg equations of motion, which follow easily from the Hamiltonian (2.12)

$$\dot{a}_{ks} = -i\omega_{ks}a_{ks} - i\sum_i g^*_{iks}(S^+_i + S^-_i), \tag{7.2}$$

$$(S^+_i + \dot{S}^-_i) = i\omega(S^+_i - S^-_i). \tag{7.3}$$

Then on combining (7.2) and (7.3) we obtain

$$\ddot{a}_{ks} + \omega^2_{ks}a_{ks} = \sum_i g^*_{ks} e^{-ik\cdot r_i}\{(\omega - \omega_{ks})S^+_i - (\omega + \omega_{ks})S^-_i\}. \tag{7.4}$$

Let $E^{(+)}$ be the positive-frequency part of the radiation field

$$E^{(+)}(r,t) = i\sum(2\pi ck/L^3)^{\frac{1}{2}}\,\varepsilon_{ks}\,e^{ik\cdot r}a_{ks}, \tag{7.5}$$

and hence in view of (7.4) it obeys the equation

$$\Box^2 E^{(+)} \equiv (\nabla^2 - c^{-2}\partial^2/\partial t^2)E^{(+)} = -(4\pi/c)J^{(+)}, \tag{7.6}$$

where

$$J^{(+)}(r,t) = (1/2L^3)\sum_{ksi} e^{ik\cdot(r-r_i)}k\varepsilon_{ks}(\varepsilon_{ks}\cdot d)[(\omega+\omega_{ks})\cdot S^-_i - S^+_i(\omega_{ks}-\omega)]. \tag{7.7}$$

The solution of (7.6) is

$$E^{(+)}(r,t) = c^{-1}\!\int d^3r_1 |r-r_1|^{-1} J^{(+)}(r_1, t-|r-r_1|/c)$$

$$= -(1/16\pi^3 c)\!\int k^3 dk\, d\Omega\, e^{ik\cdot(r_1-r_i)}\,\hat{k}\times(\hat{k}\times d)|r-r_1|^{-1} \tag{7.8}$$

$$\cdot\{(\omega+\omega_k)S^-_i(t-|r-r_1|/c) - (\omega-\omega_k)S^+_i(t-|r-r_1|/c)\}\,d^3r_1,$$

where $\hat{k}$ is the unit vector in the direction k. In obtaining (7.8) we took the limit $L^3 \to \infty$ and used the relation (2.4).

We will first calculate the field in the *radiation zone* $(r \gg (r_{ij})_{max})$ and therefore we make the standard approximation

$$|r-r_1| \approx |r| - r\cdot r'/r;$$

however, some care has to be used because of the extra phase factors $e^{-ik \cdot r_i}$. In view of this, for each i we make the approximation

$$|r - r_1| = |R_i - R_{1i}| \approx |R_i| - R_i \cdot R_{1i}/|R_i|,$$
$$R_i = r - r_i, \qquad R_{1i} = r_1 - r_i. \tag{7.9}$$

Moreover, we write the operators $S_i^{\pm}$ as

$$S_i^{\pm}(t) = S_i^{\pm I}(t) e^{\pm i\omega t}, \tag{7.10}$$

where we have separated out the unperturbed time dependence. We next make the approximation

$$S_i^{\pm I}(t - |r - r_1|/c) \approx S_i^{\pm I}(t - |r|/c). \tag{7.11}$$

On combining (7.9) – (7.11), we obtain

$$S_i^{\pm}(t - |r - r_1|/c) \approx S_i^{\pm}(t - |r|/c) \cdot \exp\{\pm ik_0 [\hat{R}_i \cdot (r_1 - r_i) + \hat{r} \cdot r_i]\}. \tag{7.12}$$

On substituting (7.12) in (7.8) and on carrying out the integrations we obtain

$$E^{(+)}(r, t) \sim E_0^{(+)}(r, t) - k_0^2 \sum_i [\hat{R}_i \times (\hat{R}_i \times d)] R_i^{-1} e^{-ik_0 \hat{r} \cdot r_i} S_i^-(t - |r|/c), \tag{7.13}$$

where we have now inserted the solution of the homogeneous part of (7.6), which is just the free-field part. The term S_i^+ does not contribute to $E^{(+)}$ because in the integrations we pick up the value only at the energy shell. Equation (7.13) is very basic as it relates the far-field-zone behavior of the field to the properties of the atomic system, which are in turn determined from the solution of the master equation. In particular, the normally ordered correlation function for the field is given by

$$\langle E^{(-)}(r, t) E^{(+)}(r', t') \rangle = k_0^4 \sum_{ij} \{\hat{R}_i \times (\hat{R}_i \times d)\} \{\hat{R}_j' \times (\hat{R}_j' \times d)\} (R_i R_j')^{-1}$$
$$\cdot e^{ik_0(\hat{r} \cdot r_i - \hat{r}' \cdot r_j)} \langle S_i^+(t - |r|/c) S_j^-(t' - |r'|/c) \rangle. \tag{7.14}$$

The free-field term does not contribute to (7.14) because the field is initially in the vacuum state. The two-time correlation function which appears in (7.14) is to be obtained from the solution of the master equation and the quantum regression theorem.

We now present an expression for the radiation rate in the far zone. What we should really caculate is the normal component of the Poynting vector [49], i.e.

$$r^2 \hat{r} \cdot S = (c/4\pi) \langle : E \times B : \rangle \cdot \hat{r} r^2 \equiv I_{\hat{r}}(t), \tag{7.15}$$

where E and B are the electric and the magnetic fields and the two colons denote the normal ordering. On using (7.13) and a similar expression for the magnetic field, it is easy to show that $I_{\hat{r}}$ is given by

$$I_{\hat{r}}(t) = (3\gamma\omega/4\pi) \sin^2\theta \sum_{ij} e^{i k_0 \hat{r} \cdot (r_i - r_j)} \langle S_i^+ (t - |r|/c) \, S_j^- (t - |r|/c) \rangle , \qquad (7.16)$$

where θ is the angle which the vector r makes with d. On summing over a small solid angle around $\hat{r}$, (7.16) reduces to

$$I(t) = \int I_{\hat{r}}(t) \, d\Omega_{\hat{r}} = 2\omega \sum_{ij} \gamma_{ij} \langle S_i^+ (t) \, S_j^- (t) \rangle , \qquad (7.17)$$

where we have ignored the retardation effect and γ_{ij} is given by (6.45). We will refer to $I(t)$ as the total radiation rate. If we compare (7.17) with (6.64), we see that

$$I(t) = -\omega(\partial/\partial t) \sum_i \langle S_i^z \rangle , \qquad (7.18)$$

i.e. the radiation rate is equal to the rate at which the atoms dissipate their energy, which is expected. This is indeed the energy conservation law of Rehler and Eberly [49].

We will now calculate the mean number of photons in any mode ks

$$N_{ks}(t) = \langle a_{ks}^+(t) a_{ks}(t) \rangle . \qquad (7.19)$$

On integrating the Heisenberg equation (7.2) we have

$$a_{ks}(t) = a_{ks}(0) \, e^{-i\omega_{ks}t} - i \sum_i g_{iks}^* \int_0^t d\tau [S_i^+(\tau) + S_i^-(\tau)] e^{-i\omega_{ks}(t-\tau)} . \qquad (7.20)$$

On substituting (7.20) in (7.19) we obtain

$$N_{ks}(t) = \sum_{ij} g_{jks}^* g_{iks} \int_0^t \int dt_1 \, dt_2 \, e^{-i\omega_{ks}(t_1 - t_2)} \\ \cdot \langle (S_i^+ (t_1) + S_i^- (t_1)) (S_j^+ (t_2) + S_j^- (t_2)) \rangle . \qquad (7.21)$$

The terms like $\langle a_{ks}^+(0) \, a_{ks}(0) \rangle$, $\langle (S_i^+ + S_i^-) \, a_{ks}(0) \rangle$ do not contribute, as the field is in vacuum state at $t = 0$. On making the RWA (7.21) reduces to[11]

$$N_{ks}(t) = \sum_{ij} g_{iks} g_{jks}^* \int_0^t \int dt_1 \, dt_2 \, e^{-i\omega_{ks}(t_1 - t_2)} \langle S_i^+ (t_1) \, S_j^- (t_2) \rangle . \qquad (7.22)$$

---

[11] As we remarked earlier, the master equation is same whichever form of the interaction is used, but the line shape is different because of the different dependence of g_{ks} on k. With $d \cdot E$ interaction we recall that $g_{ks} \propto \sqrt{k}$.

It is a simple matter to show that $N_{ks}(t)$ can be rewritten in the form

$$N_{ks}(t) = \sum_{ij} g_{iks} g_{jks}^* \int_0^t dt_1 \int_0^{t_1} dt_2 \, e^{-i\omega_{ks}(t_1 - t_2)} \langle S_i^+(t_1) S_j^-(t_2) \rangle + \text{H.C.} . \quad (7.23)$$

The rate of change of photons in the mode ks is given by

$$\sigma_{ks}(t) \equiv \dot{N}_{ks}(t) = \sum_{ij} g_{iks} g_{jks}^* \int_0^t d\tau \, e^{-i\omega_{ks}(t - \tau)} \langle S_i^+(t) S_j^-(\tau) \rangle + \text{H.C.} . \quad (7.24)$$

It is also evident that

$$\sum_{ks} \sigma_{ks}(t) = \sum_{ijks} g_{iks} g_{jks}^* \int_0^t d\tau \, e^{-i\omega_{ks}(t - \tau)} \langle S_i^+(t) S_j^-(\tau) \rangle + \text{H.C.}$$

$$\approx 2 \sum_{ij} \gamma_{ij} \langle S_i^+(t) S_j^-(t) \rangle , \quad (7.25)$$

where we made use of our Born and Markov approximations.

Let us write

$$\langle S_i^+(t_1) \rangle = \sum_\alpha f_{i\alpha}(t_1 - t_2) \langle Q_\alpha(t_2) \rangle , \quad (7.26a)$$

then by the regression theorem

$$\langle S_i^+(t_1) S_j^-(t_2) \rangle = \sum_\alpha f_{i\alpha}(t_1 - t_2) \langle Q_\alpha(t_2) S_j^-(t_2) \rangle$$

$$= \sum_\alpha f_{i\alpha}(t_1 - t_2) h_{\alpha j}(t_2) , \quad (7.26b)$$

where

$$h_{\alpha j}(t_2) = \langle Q_\alpha(t_2) S_j^-(t_2) \rangle . \quad (7.26c)$$

It is then clear from (7.26b) and (7.23) that the Laplace transform of $N_{ks}(t)$ is given by

$$\hat{N}_{ks}(z) = z^{-1} \sum_{ij\alpha} g_{iks} g_{jks}^* \hat{f}_{i\alpha}(z + i\omega_{ks}) \hat{h}_{\alpha j}(z) + \text{H.C.} . \quad (7.27)$$

The steady-state value of $N_{ks}(t)$, if it exists, is given by

$$N_{ks}(\infty) = \lim_{z \to 0^+} z \hat{N}_{ks}(z) = \sum_{ij} g_{iks} g_{jks}^* \hat{f}_{i\alpha}(0 + i\omega_{ks}) \hat{h}_{\alpha j}(0) + \text{H.C.} . \quad (7.28)$$

Similarly the Laplace transform and the steady-state value of σ_{ks} are given by

$$\hat{\sigma}_{ks}(z) = \sum_{ij\alpha} g_{iks} g_{jks}^* \hat{f}_{i\alpha}(z + i\omega_{ks}) \hat{h}_{\alpha j}(z) + \text{H.C.} , \quad (7.29)$$

$$\sigma_{ks}(\infty) = \lim_{z \to 0^+} z \hat{\sigma}_{ks}(z) . \quad (7.30)$$

High-order correlation functions involving the field operators can be calculated similarly. It may appear from (7.13) that the equal-time commutation relations between the field operator and the matter operators are violated; however, there is a sense in which the commutation relations still hold.

Finally, it should be noted that all the above results also hold for the harmonic oscillator model with $S_i^{\mp}$ replaced by a_i, a_i^+.

8. Langevin Equations Corresponding to the Master Equation (6.53) and a c-Number Description

In this section we obtain the Langevin equations[12] that are equivalent to the master Eq. (6.53). The Langevin equations are easier to interpret. Let us assume a Langevin equation for any atomic operator Q of the form [52]

$$dQ/dt = A_Q + F_Q,\tag{8.1}$$

where A_Q is the drift vector (operator) and F_Q is a random-force operator which is delta-correlated, and its mean value vanishes i.e.

$$\langle F_Q\rangle = 0,\quad \langle F_Q(t)\,F_M(t')\rangle = 2\langle D_{QM}(t)\rangle\,\delta(t-t'),\tag{8.2a}$$

$$\langle F_M(t)\,F_Q(t')\rangle = 2\langle D_{MQ}(t)\rangle\,\delta(t-t').\tag{8.2b}$$

Note that $\langle D_{QM}\rangle \neq \langle D_{MQ}\rangle$ since one is dealing with a quantum system. The diffusion coefficient is given by

$$2\langle D_{QM}\rangle = \langle(d/dt)\,QM\rangle - \langle A_Q M\rangle - \langle Q A_M\rangle,\tag{8.3a}$$

all the terms which appear on the right-hand side of (8.3) are known from the master equation. The higher-order moments of the random force can be obtained similarly (cf. [53]), for instance

$$\langle F_Q(t_1)\,F_M(t_2)\,F_N(t_3)\ldots\rangle = n!\langle D_{QMN}\ldots\rangle\,\delta(t_1-t_2)\,\delta(t_2-t_3)\ldots,\tag{8.2c}$$

where for example

$$3!\langle D_{QMN}\rangle = \langle(d/dt)\,QMN\rangle - \langle A_Q MN\rangle - \langle Q A_M N\rangle - \langle QM A_N\rangle$$
$$- \langle(D_{QM}+D_{MQ})\,N\rangle - \langle Q(D_{MN}+D_{NM})\rangle - \langle(D_{QN}+D_{NQ})\,M\rangle_{\text{order}},\tag{8.3b}$$

---

[12] For a discussion of Langevin equations in a classical context, see e.g. [50, 51].

and where in terms like $\langle D_{QN}M \rangle_{\text{order}}$ one should retain the original order. This procedure when applied to the operator $S_i^\pm$ leads to the Langevin equation

$$\dot{S}_i^\pm = (\pm i\omega - \gamma)\, S_i^\pm + 2 \sum_{j \neq i} (\gamma_{ij} \mp i\Omega_{ij})\, S_j^\pm\, S_i^z + F_i^\pm, \tag{8.4}$$

where the random force has the properties

$$\langle F_i^\pm(t) \rangle = 0, \quad \langle F_i^\pm(t)\, F_j^\mp(t') \rangle = 2\langle D_{ij}^{\pm\mp}(t) \rangle\, \delta(t - t'), \tag{8.5}$$

and where drift and diffusion are related by the second fluctuation-dissipation theorem

$$\langle D_{ij}^{+\,-} \rangle = \langle D_{ij}^{+\,+} \rangle = \langle D_{ij}^{-\,-} \rangle = 0, \tag{8.6}$$

$$2\langle D_{ij}^{-\,+} \rangle = 2\gamma\delta_{ij} + 8\langle S_i^z(t)\, S_j^z(t) \rangle\, \gamma_{ij}(1 - \delta_{ij}), \tag{8.7}$$

which is not Gaussian [53] as it is in the classical case. It should be noted that the diffusion coefficient is a function of the atomic mean values and hence the above Langevin equations are highly nonlinear. For these reasons the operator Langevin equations are not very useful for practical calculations. The first term represents the free motion of the spin, the term $-\gamma S_i^\pm$ describes the radiative decay, and the third term, which depends on the presence of other atoms, is responsible for coherence in the emission process as well as the atomic correlations. The random force preserves the atomic commutation relations. The Langevin equation for S_i^z is

$$\dot{S}_i^z = - \sum_j \gamma_{ij}(S_i^+ S_j^- + S_j^+ S_i^-) - i \sum_{j \neq i} \Omega_{ij}(S_i^+ S_j^- - S_j^+ S_i^-) + F_i^z. \tag{8.8}$$

We rewrite the Langevin equations (8.4), (8.8) in the vectorial form

$$\dot{S}_i = : \sum_j \omega_{ij} \times S_i : + F_i, \quad S_i = (S_i^x, S_i^y, S_i^z), \tag{8.9a}$$

where the vector ω, in the rotating frame, is given by

$$\omega_{ij} = \{(-2\gamma_{ij}S_j^y + 2\Omega_{ij}S_j^x), (2\gamma_{ij}S_j^x + 2\Omega_{ij}S_j^y), 0\}, \tag{8.10}$$

(with the convention that $\Omega_{ij} \equiv 0$ for the diagonal terms $i = j$, i.e. $\omega_{ii} = \{-2\gamma S_i^y, 2\gamma S_i^x, 0\}$). The two dots in (8.9) indicate the normal ordering, i.e. one should arrange all the S_i^+ (S_i^-) operators to the extreme left (extreme right) and all the S_i^z operators in the middle of S_i^+ and S_i^-.

If we adopt this prescription, (8.4) can be written as

$$\dot{S}_i^\pm = \pm i\omega S_i^\pm + 2\gamma : S_i^\pm S_i^z : + 2 \sum_{j \neq i} (\gamma_{ij_\mp} i\Omega_{ij}) S_j^\pm S_i^\pm + F_i^\pm ; \qquad (8.9b)$$

for the terms with $i \neq j$ no prescription is required since the atomic operators corresponding to the different atoms commute. Equation (8.9b) means that

$$\dot{S}_i^+ = i\omega S_i^+ + 2\gamma S_i^+ S_i^z + 2 \sum_{j \neq i} (\gamma_{ij} - i\Omega_{ij}) S_j^+ S_i^z + F_i^+ ,$$

$$\dot{S}_i^- = -i\omega S_i^- + 2\gamma S_i^z S_i^- + 2 \sum (\gamma_{ij} + i\Omega_{ij}) S_j^- S_i^z + F_i^- ,$$

and these are identical to (8.4). With this prescription we see that the Langevin equations describing spontaneous emission can be written in the form of the Bloch equations (cf. [16]) that are conventionally used to describe the atomic systems in the external (c-number) electric field. Since in the case of spontaneous emission we are dealing with a quantized electromagnetic field, we see that the vector ω is itself a functional of the atomic operators; moreover, the operator random forces appear in the Bloch equations, which makes them the operator stochastic equations.

We will now obtain another set of Langevin equations expressed in terms of the c-number variables and which contain no fluctuating forces. Such Langevin equations are therefore much easier to handle. We introduce the characteristic function defined by

$$C(\{\alpha_i\}, \{\alpha_i^*\}, t) \equiv \left\langle \prod_i e^{i\alpha_i S_i^+} e^{i\alpha_i^* S_i^-} \right\rangle \equiv \langle G \rangle . \qquad (8.11)$$

This function is useful to calculate the normally ordered expectation values of the form $\langle S_i^+ S_j^+ S_k^+ \dots S_l^- S_m^- \dots \rangle$. The equation of motion for C is obtained from (6.62)

$$\partial C/\partial t = -i\omega \langle [G, S_i^+] S_i^- - S_i^+ [S_i^-, G] \rangle - i \sum_{i \neq j} \Omega_{ij} \langle [G, S_i^+] S_j^-$$

$$- S_i^+ [S_j^-, G] \rangle - \sum_{ij} \gamma_{ij} \{ \langle [G, S_i^+] S_j^- \rangle + \langle S_i^+ [S_j^-, G] \rangle \} . \qquad (8.12)$$

To simplify (8.12) we use the identities

$$[e^{i\alpha_i^* S_i^-}, S_i^+] = -2i\alpha_i^* S_i^+ S_i^- e^{i\alpha_i^* S_i^-} + i\alpha_i^* e^{i\alpha_i^* S_i^-} - (i\alpha_i^*)^2 S_i^- e^{i\alpha_i^* S_i^-} , \qquad (8.13)$$

$$e^{i\alpha_i S_i^+} e^{i\alpha_i^* S_i^-} S_i^+ S_i^- = e^{i\alpha_i S_i^+} S_i^+ e^{i\alpha_i^* S_i^-} S_i^- + (i\alpha_i^*) e^{i\alpha_i S_i^+} e^{i\alpha_i^* S_i^-} S_i^- , \qquad (8.14)$$

and their Hermitian adjoints. A straightforward analysis shows that (cf. [31])

$$\partial C/\partial t = - i\omega \sum_i \{i\alpha_i^* \, \partial C/\partial(i\alpha_i^*) - i\alpha_i \, \partial C/\partial(i\alpha_i)\} - \gamma \sum_i \{i\alpha_i^* \, \partial C/\partial(i\alpha_i^*)$$

$$+ i\alpha_i \, \partial C/\partial(i\alpha_i)\} - \sum_{i \neq j} \left[(\gamma_{ij} + i\Omega_{ij}) \left\{ - 2i\alpha_i^* \, \frac{\partial^3 C}{\partial(i\alpha_i) \, \partial(i\alpha_i^*) \, \partial(i\alpha_j^*)} \right. \right.$$

$$+ i\alpha_i^* \, \partial C/\partial(i\alpha_j^*) - (i\alpha_i^*)^2 \, \frac{\partial^2 C}{\partial(i\alpha_i^*) \, \partial(i\alpha_j^*)} \bigg\} \tag{8.15}$$

$$\left. + \text{ terms with } \alpha \to \alpha^*, \Omega \to -\Omega \right].$$

Since the operators $S_i^\pm$ obey the relation $(S_i^\pm)^2 = 0$, (8.15) reduces to

$$\partial C/\partial t = - (i\omega + \gamma) \sum_i i\alpha_i^* \, \partial C/\partial(i\alpha_i^*) - (- i\omega + \gamma) \sum_i i\alpha_i \, \partial C/\partial(i\alpha_i)$$

$$- \sum_{i \neq j} \left[(\gamma_{ij} + i\Omega_{ij}) \left\{ - 2i\alpha_i^* \, \frac{\partial^3 C}{\partial(i\alpha_i) \, \partial(i\alpha_i^*) \, \partial(i\alpha_j^*)} + i\alpha_i^* \, \partial C/\partial(i\alpha_j^*) \right\} \right.$$

$$\left. + \text{ terms with } \alpha \to \alpha^*, \Omega \to -\Omega \right]. \tag{8.16}$$

We now introduce the distribution function $P(\{z_i\}, \{z_i^*\}, t)$ defined by

$$P(\{z_i\}, \{z_i^*\}, t) = \pi^{-2N} \int d^2\{\alpha_i\} \, C(\{\alpha_i\}, \{\alpha_i^*\}, t) \prod_i e^{-i(\alpha_i z_i^* + \alpha_i^* z_i)}. \tag{8.17}$$

On combining (8.16) and (8.17) we find that the distribution function obeys the equation

$$\partial P/\partial t = (i\omega + \gamma) \sum_i (\partial/\partial z_i) (z_i P) + \sum_{i \neq j} (\gamma_{ij} + i\Omega_{ij}) (\partial/\partial z_i)$$

$$\cdot (z_j \{1 - 2|z_i|^2\} P) + \text{c.c.} . \tag{8.18}$$

The distribution function P satisfies an equation of motion which involves only the first-order derivatives. Note that the distribution function P as defined by (8.17) is in general a highly singular object. The Eq. (8.18) is equivalent to the Langevin equations

$$\dot{z}_i = - (i\omega + \gamma) z_i + \sum_{j \neq i} (\gamma_{ij} + i\Omega_{ij}) z_j (2|z_i|^2 - 1), \quad i = 1, 2, ..., N . \tag{8.19}$$

These Langevin equations provide us with an alternative description of spontaneous emission from a collection of identical two-level atoms. No fluctuating force appears in (8.19) due to the fact that we obtained equations of motion for the normally ordered characteristic function.

Any other order will give a fluctuating force in (8.19). The absence of such a force in (8.19) makes such equations rather useful for the study of emission, although they have to be handled carefully because of their spin $-\frac{1}{2}$ characteristics. The solution of (8.19) would enable us to calculate all the normally ordered atomic expectation values and correlation functions, i.e. mean values like $\langle z_i^* z_j^* \dots z_l z_m \dots \rangle$ correspond to $\langle S_i^+ S_j^+ \dots S_l^- S_m^- \rangle$. It should further be noted that the Eqs. (8.19) are in the form of an N-dimensional Van der Pol [54] oscillator with the sign changed. We remark that such equations also occur in the theory of multimode lasers [55]. Some applications of (8.19) will be considered in Chapter 14.

The Langevin equations are very useful in obtaining the mean value equations. We now present some equations of motion. As we are considering the case of two-level atoms, we can always put terms like $z_i^2 = z_i^{*2} = 0$. It is immediately obvious from (8.19) that

$$\langle \dot{S}_i^- \rangle = -(i\omega + \gamma) \langle S_i^- \rangle + 2 \sum_{j \neq i} (\gamma_{ij} + i\Omega_{ij}) \langle (2S_i^+ S_i^- - 1) S_j^- \rangle$$

$$= -(i\omega + \gamma) \langle S_i^- \rangle + 2 \sum_{j \neq i} (\gamma_{ij} + i\Omega_{ij}) \langle S_i^z S_j^- \rangle , \qquad (8.20)$$

$$\langle \dot{z_i^* z_i} \rangle = - \sum_j \gamma_{ij} \langle z_i^* z_j + z_i z_j^* \rangle - i \sum_{j \neq i} \Omega_{ij} \langle z_i^* z_j - z_j^* z_i \rangle \Rightarrow$$

$$\langle \dot{S_i^+ S_i^-} \rangle = - \sum_j \gamma_{ij} \langle S_i^+ S_j^- + S_j^+ S_i^- \rangle - i \sum_{j \neq i} \Omega_{ij} \langle S_i^+ S_j^- - S_j^+ S_i^- \rangle , \qquad (8.21)$$

$$\langle \dot{z_i^* z_j} \rangle = -2\gamma \langle z_i^* z_j \rangle + \sum_{l \neq i} (\gamma_{il} - i\Omega_{il}) \langle z_l^* z_j (2|z_i|^2 - 1) \rangle$$

$$+ \sum_{l \neq j} (\gamma_{jl} + i\Omega_{jl}) \langle z_i z_l^* (2|z_j|^2 - 1) \rangle \Rightarrow$$

$$\langle \dot{S_i^+ S_j^-} \rangle + 2\gamma \langle S_i^+ S_j^- \rangle = 4\gamma_{ij} \langle S_i^+ S_i^- S_j^+ S_j^- \rangle - \gamma_{ij} \langle S_i^+ S_i^- + S_j^+ S_j^- \rangle$$

$$- i\Omega_{ij} \langle S_i^+ S_i^- - S_j^+ S_j^- \rangle + \sum_{l \neq j \neq i} (\gamma_{jl} + i\Omega_{jl}) \langle S_i^+ S_j^z S_l^- \rangle$$

$$+ \sum_{l \neq j \neq i} (\gamma_{il} - i\Omega_{il}) \langle S_l^+ S_i^z S_j^- \rangle , \qquad (i \neq j), \qquad (8.22)$$

$$\langle |z_i|^2 \dot{|z_j|^2} \rangle + 4\gamma \langle |z_i|^2 |z_j|^2 \rangle = - \sum_{l \neq i \neq j} \{ (\gamma_{il} + i\Omega_{il}) \langle |z_j|^2 z_i^* z_l \rangle$$

$$+ (\gamma_{il} - i\Omega_{il}) \langle |z_j|^2 z_i z_l^* \rangle + \text{terms with } i \to j \} , \qquad (i \neq j) \Rightarrow$$

$$\langle S_i^+ S_j^+ \dot{S_i^- S_j^-} \rangle + 4\gamma \langle S_i^+ S_j^+ S_i^- S_j^- \rangle = - \sum_{l \neq i \neq j} \{ (\gamma_{il} + i\Omega_{il}) \langle S_i^+ S_j^+ S_j^- S_l^- \rangle$$

$$+ (\gamma_{il} - i\Omega_{il}) \langle S_l^+ S_j^+ S_j^- S_i^- \rangle + \text{terms with } i \to j \} , \qquad (i \neq j) . \qquad (8.23)$$

Finally we also have

$$\left\langle \prod_i \dot{|z_i|^2} \right\rangle = -2N\gamma \left\langle \prod_i |z_i|^2 \right\rangle \Rightarrow$$

$$\left\langle \prod_i S_i^+(t) S_i^-(t) \right\rangle = e^{-2N\gamma t} \left\langle \prod_i S_i^+(0) S_i^-(0) \right\rangle, \tag{8.24}$$

which shows that the probability that all the atoms will remain in the excited state decays exponentially, the decay constant being N times that of a single atom. This result is independent of geometry, apart from the conditions imposed on it by (6.34) and (6.35). The mean value equations which we presented above also follow from (6.62) except that the present derivation is much simpler; note, however, that all the hard work has been done in obtaining (8.19).

For the harmonic oscillator model [master Eq. (6.65)] the Langevin equations are rather simple. On passing to the coherent-state representation [56]

$$\varrho = \int \Phi(\{z\}) |\{z\}\rangle \langle\{z\}| d^2 \{z\}, \tag{8.25}$$

where $|z\rangle$ is a coherent state which is an eigenstate of the annihilation operator and where Φ is known as the Sudarshan-Glauber distribution function, we find that Φ satisfies the equation[13]

$$\partial\Phi/\partial t = (i\omega + \gamma) \sum_i (\partial/\partial z_i)(z_i\Phi) + \sum_{i \neq j} (\gamma_{ij} + i\Omega_{ij})(\partial/\partial z_i)(z_j\Phi) + \text{c.c.} . \tag{8.26}$$

The differential Eq. (8.26) is easily integrated. The corresponding Langevin equations are

$$\dot{z}_i = -(i\omega + \gamma) z_i - \sum_{j \neq i} (\gamma_{ij} + i\Omega_{ij}) z_j . \tag{8.27}$$

The fluctuating force again does not occur in (8.27) due to our choice of the coherent-state representation. The Langevin Eqs. (8.27) are linear. If we compare (8.19) with (8.27), we see immediately how the nonlinearity appears for the case of two-level atoms. For one two-level atom there is no nonlinearity, and the Langevin equations for the two-level atom and the oscillator model are identical except for a difference in the value of the renormalized frequency shift. It should further be noted that when each atom is in a state very close to the upper state (ground state) (8.19) can be linearized by replacing $|z_i|^2$ by $+1(-1)$, and then (8.19) describes the dynamics of the emission very close to the initial instant (final instant).

---

[13] The transformation of the operator equations to c-number equations is discussed in Ref. [37].

Langevin equations such as (8.1) and (8.4) can also be obtained directly from the Heisenberg equations of motion. Indeed, this was the procedure used by Senitzky [58] in connection with simple systems and by Haken and coworkers [55, 59] (see also Lax [46, 52]) in connection with the theory of lasers. The same procedure was used by Lehmberg [34] to obtain the master Eq. (6.53). This method has also attracted a good deal of attention in connection with spontaneous emission[14] [41, 60, 61]. We now outline this method briefly.

The Heisenberg equations of motion for any arbitrary atomic operator Q and the field operators a_{ks} are

$$\mathrm{d}a_{ks}/\mathrm{d}t = -i\omega_{ks}a_{ks} - i\sum_j g^*_{jks}(S^+_j + S^-_j),$$ (8.28)

$$\mathrm{d}Q/\mathrm{d}t = -i\omega\left[Q, \sum_i S^z_i\right] - i\sum_{jks} g_{jks}[Q, S^+_j + S^-_j]a_{ks}$$

$$-i\sum_{jks} g^*_{jks}a^+_{ks}[Q, S^+_j + S^-_j].$$ (8.29)

On integrating (8.28) we have

$$a_{ks}(t) = a_{ks}(0)\,e^{-i\omega_{ks}t} - i\sum_j g^*_{jks}\int_0^t [S^+_j(\tau) + S^-_j(\tau)]\,e^{-i\omega_{ks}(t-\tau)}\,\mathrm{d}\tau.$$ (8.30)

On substituting (8.30) in (8.29) we find

$$\mathrm{d}Q/\mathrm{d}t = -i\omega[Q, \Sigma S^z_j] - i\sum_{jks} g_{jks}[Q, S^+_j + S^-_j]a_{ks}(0)\,e^{-i\omega_{ks}t}$$

$$-i\sum_{jks} g^*_{jks}a^+_{ks}(0)\,e^{i\omega_{ks}t}[Q, S^+_j + S^-_j]$$

$$-\sum_{jlks} g^*_{lks}g_{jks}[Q, S^+_j + S^-_j]\int_0^t [S^+_l(\tau) + S^-_l(\tau)]\,e^{-i\omega_{ks}(t-\tau)}\,\mathrm{d}\tau$$ (8.31)

$$+\sum_{jlks} g^*_{jks}g_{lks}\int_0^t [S^+_l(\tau) + S^-_l(\tau)]\,e^{i\omega_{ks}(t-\tau)}\,\mathrm{d}\tau[Q, S^+_j + S^-_j].$$

So far, Eq. (8.31) is exact. Note that in (8.29) we put the field annihilation operators (creation operators) to the right (left) of the atomic operators ("normal order"). This has the advantage that the mean values of the operators

$$[Q, S^+_j + S^-_j]a_{ks}(0), \qquad a^+_{ks}(0)[Q, S^+_j + S^-_j],$$

---

[14] This work on the so-called radiation-reaction theories has been done by Bullough and coworkers [60, 61] and by Eberly and coworkers [41]. Many of the results of operator radiation-reaction theories coincide with the results of the master-equation approach. For some comments on radiation reaction, see also [62].

vanish, since the electromagnetic field is in the vacuum state at $t = 0$, and this obviously *simplifies* the calculations quite a bit. Normal ordering is *by no means essential* here. Under the approximations which we used to obtain the master Eq. (6.53), (8.31) can be simplified further. The approximations (6.34) and (6.35) and the limit (6.33) essentially enable one to replace the operators in the integrand by

$$S_j^{\pm}(t - \tau) \approx S_j^{\pm}(t)\, e^{\mp i\omega\tau}\,.$$

In addition, on making RWA (8.31) reduces to (for details see Lehmberg [34])

$$\mathrm{d}Q/\mathrm{d}t = i(\omega + \Omega_{ii})\Sigma[S_j^z, Q] + i\sum_{i \neq j}\Omega_{ij}[S_i^+ S_j^-, Q] - \sum_{ij}\gamma_{ij}$$

$$\cdot\{S_i^+ S_j^- Q - 2S_i^+ Q S_j^- + Q S_i^+ S_j^-\} + F_Q\,, \qquad (8.32)$$

where

$$F_Q = -i\sum_{jks}g_{jks}[Q, S_j^+ + S_j^-]\, a_{ks}(0)\, e^{-i\omega_{ks}t}$$

$$-i\sum_{jks}g_{jks}^* a_{ks}^+(0)\, e^{i\omega_{ks}t}[Q, S_j^+ + S_j^-]\,, \qquad (8.33)$$

and where Ω_{ii}, Ω_{ij} and γ_{ij} are given by (6.37)–(6.39). The operator force F_Q has the property

$$\langle F_Q \rangle = 0\,, \qquad (8.34)$$

and hence

$$\mathrm{d}\langle Q\rangle/\mathrm{d}t = i\sum_{i}(\omega + \Omega_{ii})\langle[S_i^z, Q]\rangle + i\sum_{i \neq j}\Omega_{ij}\langle[S_i^+ S_j^-, Q]\rangle - \sum_{ij}\gamma_{ij}$$

$$\cdot\{\langle S_i^+ S_j^- Q - 2S_i^+ Q S_j^- + Q S_i^+ S_j^-\rangle\}\,, \qquad (8.35)$$

which is seen to be equivalent to the master Eq. (6.53), since Q is an arbitrary atomic operator. From (8.32) it is clear that all the reference to the field operators appears in the random force F_Q. It is apparent from the above derivation that we did not make any approximation on the random force. These approximations will enter into the calculation of the correlation functions of the random force. If, for example, we take for Q the operator S_i^+, then (8.32) will coincide with (8.4) and with the random force equal to

$$F_i^+(t) = -2i\sum_{ks}g_{iks}^* a_{ks}^+(0)\, e^{i\omega_{ks}t}\, S_i^z(t)\,, \qquad (8.36)$$

where we have ignored the other term in view of RWA. It is clear from (8.36) that

$$\langle F_i^+(t) F_j^-(t')\rangle = \langle F_i^+(t) F_j^+(t')\rangle = \langle F_i^-(t) F_j^-(t')\rangle = 0,$$

$$\langle F_i^-(t) F_j^+(t')\rangle = 4 \sum_{k_1 s_1 k_2 s_2} g_{ik_1 s_1} g^*_{jk_2 s_2} e^{-i\omega_{k_1 s_1}t}$$

$$\cdot e^{i\omega_{k_2 s_2}t'} \langle S_i^z(t) a_{k_1 s_1} a^+_{k_2 s_2} S_j^z(t')\rangle .$$

$$(8.37)$$

Equation (8.37) is not particularly convenient for calculating the correlation function, although such equations have been used by Haken and others (see also [55], pp. 43, 51) to compute the correlation function under the same kind of approximations which enable one to make Born and Markov approximations. Under these approximations (8.37) reduces to (8.7). There is an associated problem, that of the commutation relations. It can be shown that the Langevin Eq. (8.4) is such that the equal-time atomic commutation relations are preserved; if they were not, then there would be an internal inconsistency in the theory. In the master-equation approach such a problem does not seem to arise, as all the operators are in the Schrödinger picture.

Because of linearity it is much easier to calculate the correlation functions of the random force for the harmonic oscillator model. For this model the Langevin equation is given by (8.32) (exept for the difference in the value of the renormalized frequency) with the random force F_i^- equal to

$$F_i^- = -i\Sigma g_{iks} a_{ks}(0) e^{-i\omega_{ks}t}, \qquad (8.38)$$

and therefore in the Markov approximation

$$\langle F_i^-(t) F_j^+(t')\rangle = 2\langle D_{ij}^{-+}\rangle \delta(t-t'),$$

$$2\langle D_{ij}^{-+}\rangle = 2\gamma_{ij}.$$

$$(8.39)$$

We now turn to the calculation of the electric field at the position of the ith dipole. On combining (7.5) and (8.30) we obtain for the positive frequency part of the electric-field operator

$$E^{(+)}(r_i, t) = E_0^+(r_i, t) + E_{ii}^+(r_i, t) + \sum_{j\neq i} E_{ij}^{(+)}(t), \qquad (8.40)$$

where

$$E_{ii}^{(+)}(t) = i\sum_{ks}(2\pi ck/L^3)(d\cdot \varepsilon^*_{ks})\,\varepsilon_{ks}\int_0^t p_i(t-\tau)e^{-ikc\tau}d\tau, \qquad (8.41)$$

$$E_{ij}^{(+)}(t) = i\sum_{ks}(2\pi ck/L^3)(d\cdot \varepsilon^*_{ks})\,\varepsilon_{ks}\,e^{ik\cdot r_{ij}}\int_0^t p_j(t-\tau)e^{-ikc\tau}d\tau, \qquad (8.42)$$

where we have put for the sake of brevity

$$p_i(t) = S_i^+(t) + S_i^-(t) , \qquad r_{ij} = r_i - r_j . \tag{8.43}$$

In (8.40) $E_{ii}^{(+)}(t)$ is the so called "operator radiation-reaction field" (we discuss this more fully in § 16) and $E_{ij}^{(+)}(t)$ is the "operator dipole field" which we consider here in detail. The total field operator $E_{ij}(t)$ is given by (we let $L^3 \to \infty$)

$$\begin{aligned}
E_{ij}(t) &= E_{ij}^{(+)}(t) + E_{ij}^{(-)}(t) \\
&= V \times V \times d(ic/4\pi^2) \int (d^3 k/k) \, e^{i\mathbf{k} \cdot \mathbf{r}_{ij}} \int_0^t (e^{-ikc\tau} - e^{ikc\tau}) \, p_j(t - \tau) \, d\tau \\
&= V \times V \times d(c/r_{ij}) \int_0^t p_j(t - \tau) \, \{\delta(r_{ij} - c\tau) - \delta(r_{ij} + c\tau)\} \, d\tau \qquad (8.44) \\
&= V \times V \times d p_j(t - r_{ij}/c)/r_{ij} \\
&= V \times V \times d[p_j]/r_{ij} ,
\end{aligned}$$

where $[p_j]$ refers to the retarded values. Note that (8.44) coincides with the field at the point i of a radiating dipole at the point j [cf. [63], p. 8, Eqs. (49) and (50)], except for the fact that p_j is an operator in the present case. (8.44) can be reduced in the well-known way to the form

$$\begin{aligned}
E_{ij}(t) &= (3[p_j]/r_{ij}^3 + 3[\dot{p}_j]/cr_{ij}^2 + [\ddot{p}_j]/c^2 r_{ij})(\mathbf{d} \cdot \hat{\mathbf{r}}_{ij}) \, \hat{\mathbf{r}}_{ij} \\
&\quad - ([p_j]/r_{ij}^3 + [\dot{p}_j]/cr_{ij}^2 + [\ddot{p}_j]/c^2 r_{ij}) \, \mathbf{d} .
\end{aligned} \tag{8.45}$$

If we write

$$p_j(t) = S_j^{+I}(t) \, e^{i\omega t} + S_j^{-I}(t) \, e^{-i\omega t} , \tag{8.46}$$

where $S_j^{\pm I}(t)$ are the slowly varying operators, then we can write approximately (ignoring retardation)

$$[p_j] \approx S_j^{+I}(t) \, e^{i\omega t - ik_0 r_{ij}} + \text{H.C.} , \qquad [\ddot{p}_j] = -\omega^2 [p_j] , $$
$$[\dot{p}_j] \approx i\omega \{S_j^{+I}(t) \, e^{i\omega t - ik_0 r_{ij}} - \text{H.C.}\} . \tag{8.47}$$

We substitute (8.47) in (8.45) and obtain for the dipole field

$$\begin{aligned}
E_{ij}(t) &\approx e^{ik_0 r_{ij}} S_j^- [(3/r_{ij}^3 - 3ik_0/r_{ij}^2 - k_0^2/r_{ij})(\mathbf{d} \cdot \hat{\mathbf{r}}_{ij}) \, \hat{\mathbf{r}}_{ij} \\
&\quad - (1/r_{ij}^3 - ik_0/r_{ij}^2 - k_0^2/r_{ij}) \, \mathbf{d}] + \text{H.C.} ,
\end{aligned} \tag{8.48a}$$

and therefore on using (6.45) and (6.52) we have

$$\mathbf{d} \cdot E_{ij}(t) \approx i\gamma_{ij}(S_j^-(t) - S_j^+(t)) - \Omega_{ij}(S_j^+(t) + S_j^-(t)) . \tag{8.48b}$$

In view of (8.48 b) it is interesting to note that the term $2 \sum_{j \neq i} (\gamma_{ij} + i\Omega_{ij})$ $\cdot S_j^- S_i^z$ in the Langevin Eq. (8.4) can be obtained directly even from classical electrodynamic considerations provided one everywhere replaces the classical dipole field by (8.48) [cf. Eq. (16.41)]. It would be erroneous to conclude from (8.48 b) that the positive- and negative-frequency parts of the dipole field are given by

$$\boldsymbol{d} \cdot \boldsymbol{E}_{ij}^{(\pm)}(t) = \pm i(\gamma_{ij} \pm i\Omega_{ij}) \, S_j^{\mp}(t) \,. \tag{8.49}$$

The reason is that $E_{ij}^{(+)}(t)$ also contains contributions from negative frequencies, i.e. it is not an analytic signal. A long calculation shows that (8.42) in the Markov approximation reduces to

$$\boldsymbol{d} \cdot \boldsymbol{E}_{ij}^{(+)}(t) = i\gamma_{ij} S_j^-(t) - \tfrac{1}{2}\Omega_{ij}(S_j^+(t) + S_j^-(t))$$
$$- \pi^{-1} \boldsymbol{d} \cdot \boldsymbol{V} \times \boldsymbol{V} \times (\boldsymbol{d}/r_{ij}) \psi(k_0 r_{ij}) \, \{S_j^+(t) - S_j^-(t)\} \,, \tag{8.50}$$

where the function ψ is defined by (6.71 b).

We close this section by mentioning something which is all too familiar: we obtain the equations of motion of the "mean-field theory" (cf. [64]). We restrict to the case of a single atom. From (8.29) we have on taking the mean value

$$\mathrm{d}\langle Q \rangle/\mathrm{d}t = -i\omega\langle[Q, S^z]\rangle - i \sum_{ks} g_{ks}\langle a_{ks}\rangle \, \langle[Q, S^+ + S^-]\rangle$$
$$- i \sum_{ks} g_{ks}^*\langle a_{ks}^+\rangle \, \langle[Q, S^+ + S^-]\rangle \,, \tag{8.51}$$

where we introduced the mean-field approximation, i.e. ignored all the correlations between matter and field. From (8.30) we have

$$\sum_{ks} g_{ks}\langle a_{ks}\rangle = -i \sum_{ks} |g_{ks}|^2 \int_0^t d\tau \, e^{-i\omega_{ks}\tau}\{S^+(t-\tau) + S^-(t-\tau)\} \tag{8.52}$$
$$\approx -i \sum_{ks} |g_{ks}|^2 [-i(\omega + \omega_{ks})^{-1} \, S^+ + i(\omega - \omega_{ks})^{-1} \, S^- + \pi\delta(\omega - \omega_{ks}) \, S^-] \,,$$

where we have used the Born and Markov approximations and taken the long time limit. Next we substitute (8.52) in (8.51) and make the rotating-wave approximation. This leads us to the equations of motion of the mean-field theory, e.g. we find

$$(\mathrm{d}/\mathrm{d}t) \langle S^+ \rangle = i\omega\langle S^+ \rangle + 2\gamma\langle S^z \rangle \langle S^+ \rangle - 2i\Omega^{(-)}\langle S^z \rangle \langle S^+ \rangle \tag{8.53}$$

with $\Omega^{(-)}$ defined by (6.66). The equations so obtained coincide with the equations of neoclassical theory (Chapter 16). There is, however, no *a priori* reason to ignore the matter-field correlations. We discuss the mean-field approximation more fully in Chapter 16 (see also Chapter 13).

In obtaining (8.32) we used a specialized method. There is another method due to Mori [65] which enables one to cast Heisenberg equations of motion into a Langevin equation. Mori's method as it is usually formulated applies only to equilibrium situations, whereas the spontaneous emission is essentially a nonequilibrium situation. However, it is easy to generalize Mori's method to such nonequilibrium situations; we discuss this in detail in Appendix B.

9. Perturbative Results

Before we examine the dynamical aspects of the spontaneous emission from a collection of identical two-level atoms, let us consider the results obtained from ordinary perturbation theory and derive the results of Dicke and their extensions. It is interesting that the master equation can itself be used to discuss the perturbative results because implicitly Fermi's Golden Rule has already been used in deriving the master equation.

A) Small Samples

The perturbative result for the radiation rate is given by (7.17):

$$I_0 = 2\omega \sum_{ij} \gamma_{ij} \langle S_i^+ S_j^- \rangle_0 , \tag{9.1}$$

where $\langle \ \rangle_0$ refers to the expectation value with respect to the initial density operator. We first consider the case when the atomic system is confined to a region smaller than a wavelength. On using (6.56) we find that (9.1) reduces to

$$I_0 = 2\omega\gamma \sum_{ij} \langle S_i^+ S_j^- \rangle_0 = 2\omega\gamma \langle S^+ S^- \rangle_0 , \tag{9.2}$$

where $S^\pm$ are the collective operators defined by (2.28). For the case when the atomic system was initially excited to a Dicke state $|S, m\rangle$ [Eq. (2.30)], I_0 is given by

$$I_0 = 2\omega\gamma(S + m)(S - m + 1) , \tag{9.3}$$

which shows that for $S = \frac{1}{2}N$, $m = 0$ (N even) the radiation rate is proportional to the square of the number of atoms, a phenomenon that has been referred to as "superradiance". The case when the atomic system has been prepared in a Θ state (e.g. by excitation by an external pulse) defined by

$$|\theta_0, \varphi_0\rangle_i = \cos(\theta_0/2)\, e^{i\varphi_0/2}\, |2\rangle_i + \sin(\theta_0/2)\, e^{-i\varphi_0/2}\, |1\rangle_i,$$

$$\varrho(0) = \prod_i |\theta_0, \varphi_0\rangle_i{}_i\langle\theta_0, \varphi_0|, \tag{9.4}$$

then

$$I_0 = 2\gamma\omega N \sin^2(\theta_0/2)\{1 + (N-1)\cos^2(\theta_0/2)\}, \tag{9.5}$$

which shows that the radiation rate is proportional to the square of the number of atoms when $\theta_0 = \frac{1}{2}\pi$. For $\theta_0 = \pi$ one has the normal (in-coherent) rate. Thus, the system excited initially to the state (9.4) with $\theta_0 = \frac{1}{2}\pi$ gives rise to the superradiant emission. Although the radiation rate with the states $|\frac{1}{2}N, 0\rangle$ and $|\frac{1}{2}\pi, \varphi_0\rangle$ is more or less identical, it should be borne in mind that there is a basic difference between the two cases. This is apparent if we examine the value of the dipole moment in the state $|\frac{1}{2}N, 0\rangle$ and $|\frac{1}{2}\pi, \varphi_0\rangle$. It is easily seen that

$$\langle\tfrac{1}{2}N, 0|S^+|\tfrac{1}{2}N, 0\rangle = 0, \quad \prod_i \langle\tfrac{1}{2}\pi, \varphi_0|S^+|\tfrac{1}{2}\pi, \varphi_0\rangle_i = \tfrac{1}{2}N\, e^{i\varphi_0}, \tag{9.6}$$

i.e. the dipole moment in the Dicke state vanishes whereas it has its maximum value for the Θ state with $\theta_0 = \frac{1}{2}\pi$. Moreover, in the Θ state there are no atomic correlations because it is a product state, whereas in the Dicke state the atomic correlations are quite important, for instance

$$\langle\tfrac{1}{2}N, 0|\Delta S_i^+ \Delta S_j^-|\tfrac{1}{2}N, 0\rangle \approx \tfrac{1}{4}, \quad \langle\tfrac{1}{2}\pi, \varphi_0|\Delta S_i^+ \Delta S_j^-|\tfrac{1}{2}\pi, \varphi_0\rangle = 0, \tag{9.7}$$

where $\Delta G = G - \langle G\rangle$. It is clear that the superradiance of the Dicke state can be attributed to the atomic correlations whereas the super-radiance of the Θ state is due to the existence of a macroscopic dipole moment. Because of this the present author [66] has referred to the superradiance of the Θ state and the Dicke state as being of first kind and second kind, respectively.

B) Large Samples

In the above analysis we restricted ourselves to small samples. We now consider the generalizations to the case of large systems. We first calculate the transition probability per unit time that the system will make a transition from the Dicke state $|S_1 m_1 \alpha_1\rangle$ to $|S_2 m_2 \alpha_2\rangle$, where α indicates the degeneracy. We denote this probability by $P(S_1 m_1 \alpha_1 \rightarrow S_2 m_2 \alpha_2)$.

It is evident from the master equation that this is given by

$$p(S_1 m_1 \alpha_1 \rightarrow S_2 m_2 \alpha_2)$$
$$= 2 \sum_{ij} \gamma_{ij} \langle S_2 m_2 \alpha_2 | S_j^- | S_1 m_1 \alpha_1 \rangle \langle S_1 m_1 \alpha_1 | S_i^+ | S_2 m_2 \alpha_2 \rangle . \tag{9.8}$$

The matrix elements can be calculated by using the Wigner-Eckart theorem. We have the selection rules

$$S_2 - S_1 = 0, \pm 1 , \qquad m_2 - m_1 = -1 . \tag{9.9}$$

Transitions $S_2 - S_1 = 0, S_1 = \frac{1}{2}N$

We first consider the transitions $\Delta S = 0$, with the cooperation number having the value $\frac{1}{2}N$. For $S = \frac{1}{2}N$, there is no degeneracy so we can suppress the indices α's. On using the Wigner-Eckart theorem (9.8) reduces to

$$p(\tfrac{1}{2}N, m \rightarrow \tfrac{1}{2}N, m-1) = 2(\tfrac{1}{2}N + m)(\tfrac{1}{2}N - m + 1) \sum_{ij} \gamma_{ij} T_{\frac{1}{2}N, \frac{1}{2}N}^{(j)} T_{\frac{1}{2}N, \frac{1}{2}N}^{(i)} , \tag{9.10}$$

where $T_{\frac{1}{2}N, \frac{1}{2}N}^{(i)}$ are the reduced matrix elements and are found to be[15]

$$T_{\frac{1}{2}N, \frac{1}{2}N}^{(i)} = 1/N . \tag{9.11}$$

The transition probability now becomes

$$p(\tfrac{1}{2}N, m \rightarrow \tfrac{1}{2}N, m-1) = 2N^{-2}(\tfrac{1}{2}N + m)(\tfrac{1}{2}N - m + 1) \sum_{ij} \gamma_{ij} \tag{9.12}$$

$$= 2N^{-2}(\tfrac{1}{2}N + m)(\tfrac{1}{2}N - m + 1) \left\{ N\gamma + \sum_{i \neq j} \gamma_{ij} \right\} = p_{\text{incoh}} + p_{\text{coh}} , \tag{9.13}$$

where

$$p_{\text{incoh}} = 2N^{-1}(\tfrac{1}{2}N + m)(\tfrac{1}{2}N - m + 1) \gamma , \tag{9.14}$$

$$p_{\text{coh}} = 2(\tfrac{1}{2}N + m)(\tfrac{1}{2}N - m + 1) v\gamma , \tag{9.15}$$

where

$$v = (1/\gamma N^2) \sum_{i \neq j} \gamma_{ij} . \tag{9.16}$$

---

[15] The calculation of these matrix elements is rather lengthy. To calculate these we can use the generalization of Slater's method [67]. We use the fact that $S_i^\pm, S_i^z$ for each i are tensors of rank 1 with respect to the collective operator $S = \sum_i S_i$ and use the relations

$$[S_i^+, S_j^-] = 2S_i^z \delta_{ij}, \qquad S_i^+ S_i^- + S_i^- S_i^+ = 1, \qquad S_i^+ S_i^+ = S_i^- S_i^- = 0$$

and take the matrix element of each side in terms of Dicke states (the eigenstates of S^2 and S^z). This leads to a series of recursion relations between the reduced matrix elements. The matrix elements $T_{\frac{1}{2}N, \frac{1}{2}N}^{(i)}, T_{\frac{1}{2}N, \frac{1}{2}N-1, \alpha}^{(j)}$ can also be computed directly by using the explicit formulas (2.32) and (2.33) for the state $|\tfrac{1}{2}N, \tfrac{1}{2}N\rangle$ and $|\tfrac{1}{2}N - 1, \tfrac{1}{2}N - 1, \alpha\rangle$.

The transition probability depends on v which in turn depends on the position of the atoms. We assume that the atoms are randomly distributed in a volume V. On taking the ensemble average and using (6.42) we find that v can be written as

$$v = (3(N-1)/8\pi N V) \iint_V d^3 r \, d\Omega \, e^{i k \cdot r} \{1 - (\hat{d} \cdot \hat{k})^2\} . \tag{9.17}$$

For a needle-shaped sample of cross section A and length h,

$$v = 3\pi/4 h k_0 \tag{9.18}$$

and hence

$$p_{\text{coh}} = \gamma (\tfrac{1}{2} N + m) (\tfrac{1}{2} N - m + 1) (3\pi/2 h k_0) . \tag{9.19}$$

Therefore the radiation rate is considerably reduced compared to the case of small samples (when $v = 1 - 1/N$). The parameter is almost universal to the treatment of large samples [49, 29, 70, 71].

Transitions $S_2 - S_1 = -1, S_1 = \tfrac{1}{2} N$

The states $|\tfrac{1}{2} N - 1, m\rangle$ are degenerate, the degeneracy of the state being equal to $(N-1)$ [cf. (2.31)]. From the nature of the Hamiltonian it is also clear that the probability of transition to any of these degenerate states would be the same; hence (9.8) can be written as

$$p(\tfrac{1}{2} N, m \to \tfrac{1}{2} N - 1, m - 1, \alpha) = 2(N-1)^{-1} \sum_{i j \alpha} \gamma_{ij} \langle \tfrac{1}{2} N - 1, m - 1, \alpha | S_j^- |$$
$$\tfrac{1}{2} N, m \rangle \langle \tfrac{1}{2} N, m | S_i^+ | \tfrac{1}{2} N - 1, m - 1, \alpha \rangle . \tag{9.20}$$

On using the Wigner-Eckart theorem (9.20) reduces to

$$p(\tfrac{1}{2} N, m \to \tfrac{1}{2} N - 1, m - 1, \alpha) = 2(N-1)^{-1} (\tfrac{1}{2} N + m) (\tfrac{1}{2} N + m - 1) \sum_{ij} \gamma_{ij}$$
$$\cdot \sum_\alpha T^{(i)}_{\frac{1}{2} N, \frac{1}{2} N - 1, \alpha} T^{(j)*}_{\frac{1}{2} N, \frac{1}{2} N - 1, \alpha} , \tag{9.21}$$

where T's are the reduced matrix elements. By a long calculation one finds that (9.21) reduces to

$$p(\tfrac{1}{2} N, m \to \tfrac{1}{2} N - 1, m - 1, \alpha) = 2(N-1)^{-2} (\tfrac{1}{2} N + m) (\tfrac{1}{2} N + m - 1)$$
$$\cdot \left\{ \gamma(1 - 1/N) - N^{-2} \sum_{i \neq j} \gamma_{ij} \right\} \tag{9.22}$$

$$= 2(N-1)^{-2} (\tfrac{1}{2} N + m) (\tfrac{1}{2} N + m - 1) \gamma \{1 - 1/N - v\} , \tag{9.23}$$

where we used (9.16). For small samples the above transition rate is zero as $v = (1 - 1/N)$.

For the case of a completely inverted system $m = \frac{1}{2}N$

$$p(\tfrac{1}{2}N, \tfrac{1}{2}N \to \tfrac{1}{2}N, \tfrac{1}{2}N - 1) = 2\gamma(1 + vN), \qquad (9.24a)$$

$$p(\tfrac{1}{2}N, \tfrac{1}{2}N \to \tfrac{1}{2}N - 1, \tfrac{1}{2}N - 1, \alpha) = 2\gamma\{1 - vN/(N-1)\}. \qquad (9.24b)$$

The total transition rate to the states with $S = \frac{1}{2}N - 1$ is

$$\sum_\alpha p(\tfrac{1}{2}N, \tfrac{1}{2}N \to \tfrac{1}{2}N - 1, \tfrac{1}{2}N - 1, \alpha) = 2\gamma\{N - 1 - vN\}. \qquad (9.24c)$$

The above results show that when $vN \ll 1$ the system tends to change the cooperation number S.

The total radiation rate for the Dicke state $|\tfrac{1}{2}N, m\rangle$ will be given by the sum of (9.13) and (9.23) (summed over all degenerate states):

$$I_0 = 2\omega\gamma(\tfrac{1}{2}N + m)N^{-1}\{(\tfrac{1}{2}N - m + 1)(1 + vN) + (\tfrac{1}{2}N + m - 1) \qquad (9.25)$$
$$\cdot (1 - Nv/(N-1))\}$$

$$= 2\omega\gamma N \quad \text{for} \quad m = \tfrac{1}{2}N, \qquad (9.26a)$$

$$= \omega\gamma N\{1 + v/2[N + 2 - (N-2)/(N-1)]\} \quad \text{for} \quad m = 0$$

$$\approx \omega\gamma N\{1 + Nv/2\} \quad \text{for large} \quad N, \quad m = 0. \qquad (9.26b)$$

The radiation rate for the Dicke state $|\tfrac{1}{2}N, 0\rangle$ is considerably reduced compared to the case of small samples because vN can be much smaller than unity.

In considerations of the transition probabilities the dipole-dipole interaction term $\Omega_{ij} S_i^+ S_j^-$ made no contribution. This term certainly affects off-diagonal elements and is responsible for the cooperative frequency shifts[16] which are due to the presence of other atoms. Let us denote by $\Delta(S, m)$ the cooperative frequency shift of the Dicke state $|S, m\rangle$. Evidently from the master equation it is given by

$$\Delta(\tfrac{1}{2}N, m) = \left\langle \tfrac{1}{2}N, m \left| \sum_{i \neq j} \Omega_{ij} S_i^+ S_j^- \right| \tfrac{1}{2}N, m \right\rangle$$

$$= \sum_{i \neq j} \Omega_{ij} \langle \tfrac{1}{2}N, m | S_i^+ | \tfrac{1}{2}N, m - 1 \rangle \langle \tfrac{1}{2}N, m - 1 | S_j^- | \tfrac{1}{2}N, m \rangle \qquad (9.27)$$

$$+ \sum_{i \neq j} \Omega_{ij} \sum_\alpha \langle \tfrac{1}{2}N, m | S_i^+ | \tfrac{1}{2}N - 1, m - 1, \alpha \rangle \langle \tfrac{1}{2}N - 1, m - 1, \alpha | S_j^- | \tfrac{1}{2}N, m \rangle,$$

which on using the Wigner-Eckart theorem reduces to

$$\Delta(\tfrac{1}{2}N, m) = \sum_{i \neq j} \Omega_{ij}(\tfrac{1}{2}N + m)(\tfrac{1}{2}N - m + 1) T^{(i)}_{\frac{1}{2}N, \frac{1}{2}N} T^{(j)}_{\frac{1}{2}N, \frac{1}{2}N}$$

$$+ \sum_{i \neq j} \sum_\alpha \Omega_{ij}(\tfrac{1}{2}N + m)(\tfrac{1}{2}N + m - 1) T^{(i)}_{\frac{1}{2}N, \frac{1}{2}N - 1, \alpha} T^{(j)*}_{\frac{1}{2}N, \frac{1}{2}N - 1, \alpha},$$

---

[16] For a recent review article on cooperative frequency shifts, see [68].

and this on simplification reduces to

$$\Delta(\tfrac{1}{2}N, m) = \left\{ \frac{(\tfrac{1}{2}N)^2 - m^2}{N(N-1)} \right\} \sum_{i \neq j} \Omega_{ij}. \tag{9.28}$$

The cooperative shift as given by (9.28) is symmetric in m and vanishes for the fully excited ($m = \tfrac{1}{2}N$) and the ground state. The ensemble average of (9.28) is equal to

$$\Delta(\tfrac{1}{2}N, m) = \{(\tfrac{1}{2}N)^2 - m^2\} \, V^{-2} \int_V d^3r \int_V d^3r' \, \Omega(r - r'); \tag{9.29}$$

since $\Omega(r)$ diverges as $r \to 0$, the double integrals in (9.29) are not properly defined. As is conventional, we define (9.29) as a conditionally convergent integral by extracting a small sphere about r (volume v_0) and then (9.29) reduces to

$$\Delta(\tfrac{1}{2}N, m) = ((\tfrac{1}{4}N^2 - m^2)/V^2) \int_V d^3r \int_{V-v_0} d^3r' \, \Omega(r - r'). \tag{9.30}$$

A straightforward calculation shows that (9.30) can be written as

$$\Delta(\tfrac{1}{2}N, m) = - ((\tfrac{1}{4}N^2 - m^2)/V^2)(\tfrac{3}{2}\gamma) \int_V d^3r \int_{V-v_0} d^3r' \{1 + \partial^2/\partial(k_0 z)^2\}$$
$$\cdot \frac{\cos k_0 |r - r'|}{|r - r'|} \tag{9.31}$$

$$= - ((\tfrac{1}{4}N^2 - m^2)/V^2)(\tfrac{3}{2}\gamma)\left\{ \tfrac{4}{3}\pi V + \int d^3r\{1 + \partial^2/\partial(k_0 z)^2\}\int d^3r' \frac{\cos k_0 |r - r'|}{|r - r'|} \right\}$$

$$= - \tfrac{3}{2}\gamma\{(\tfrac{1}{2}N)^2 - m^2\} \left\{ \tfrac{4}{3}\pi/V + V^{-2} \int d^3r\{1 + \partial^2/\partial(k_0 z)^2\} \right. \tag{9.32}$$
$$\left. \cdot \int d^3r' \frac{\cos k_0 |r - r'|}{|r - r'|} \right\}.$$

The double integral in (9.32) is now well defined. The first term is the analog of the Lorentz field term that has recently attracted some attention in connection with pulse propagation problems [69]. The second term is geometry-dependent. For the evaluation of such terms and other details we refer to Refs. [68, 71]. Some of the results of the present section were first derived in [71] by the application of first-order perturbation theory rather than the master equation. The fact that the perturbative results from the two approaches are identical gives us confidence in the projection operator (6.14).

10. Spontaneous Emission from a Single Two-Level Atom

In this section we consider the dynamical aspects of spontaneous emission from a single two-level atom. The master equation (6.53) reduces in this case (ignoring the suffix i) to

$$\partial\varrho/\partial t = -i\omega_0[S^z,\varrho] - \gamma(S^+S^-\varrho - 2S^-\varrho S^+ + \varrho S^+S^-), \tag{10.1}$$

where from (6.46) and (6.48) we have

$$\omega_0 = \omega - (\gamma/\pi)\ln\{|\omega_c/\omega - 1| |\omega_c/\omega + 1|\}, \qquad \gamma = \tfrac{2}{3}|d|^2\omega^3/c^3. \tag{10.2}$$

It should be noted that the coefficient γ is equal to half the Einstein A coefficient. Since ω_0 is the transition frequency, it contains the Lamb shift of both the ground state and the excited state. From (10.1) the equations of motion for the mean values are

$$\partial/\partial t\langle S^\pm\rangle = (\pm i\omega_0 - \gamma)\langle S^\pm\rangle, \qquad \partial/\partial t\langle S^+S^-\rangle = -2\gamma\langle S^+S^-\rangle, \tag{10.3}$$

which are easily integrated to

$$\langle S^\pm(t)\rangle = e^{(\pm i\omega_0 - \gamma)t}\langle S^\pm(0)\rangle, \langle S^+(t)S^-(t)\rangle = e^{-2\gamma t}\langle S^+(0)S^-(0)\rangle, \tag{10.4}$$

showing that the dipole moment decays exponentially. Moreover, the probability that the atom remains in the excited state ($\langle S^+S^-\rangle$) decreases exponentially. These results are identical to the ones from the exponential-decay theory [cf. (3.4), (3.11)].

The two-time correlation function is given by

$$\begin{aligned}\langle S^+(t)S^-(t')\rangle &= \exp\{(i\omega_0 - \gamma)(t - t')\}\langle S^+(t')S^-(t')\rangle \\ &= \exp\{i\omega_0(t - t') - \gamma(t + t')\}\langle S^+(0)S^-(0)\rangle, \qquad t > t',\end{aligned} \tag{10.5}$$

where we used (10.4) and the regression theorem (6.60). The time dependence of the density operator is given by (10.4) and

$$\begin{aligned}\varrho(t) &= \tfrac{1}{2} + 2S^z\langle S^z(t)\rangle + S^+\langle S^-(t)\rangle + S^-\langle S^+(t)\rangle; \\ \langle S^z\rangle &= -\tfrac{1}{2} + \langle S^+S^-\rangle.\end{aligned} \tag{10.6}$$

In the special case when the atom was initially in the excited state

$$\begin{aligned}&\langle S^\pm(0)\rangle = 0, \qquad \langle S^+(0)S^-(0)\rangle = 1, \\ &\varrho(t) = \tfrac{1}{2} + (2e^{-2\gamma t} - 1)S^z,\end{aligned} \tag{10.7a}$$

which implies that

$$\varrho^2(t) \neq \varrho(t) \quad \text{unless} \quad t = 0, \infty. \tag{10.7b}$$

(10.7) shows that as a result of spontaneous emission the atom is not left in the pure state. The steady-state behavior is

$$\varrho(\infty) = \tfrac{1}{2} - S^z = |2\rangle\langle 2|, \tag{10.8}$$

which is as expected.

The mean number of photons in the mode ks is obtained from (7.23) and (10.5)

$$N_{ks}(t) = |g_{ks}|^2 \int\limits_0^t dt_1 \int\limits_0^{t_1} dt_2 \, e^{-i(\omega_{ks}-\omega_0)(t_1-t_2)-\gamma(t_1+t_2)} \langle S^+(0) S^-(0) \rangle + \text{H.C.}$$

$$= |g_{ks}|^2 \{(\omega_{ks}-\omega_0)^2 + \gamma^2\}^{-1} \{1 + e^{-2\gamma t} - 2e^{-\gamma t}\cos(\omega_{ks}-\omega_0)t\}$$
$$\cdot \langle S^+(0) S^-(0) \rangle . \tag{10.9}$$

Let $p_{ks}(t)$ be the probability that there be a photon in the mode ks. $p_{ks}(t)$ must be identical to $N_{ks}(t)$ since in our problem there is at best one photon or, to be more explicit, we can write

$$p_{ks}(t) = \langle |\{ks\}\rangle \langle\{ks\}| \rangle_t = \langle a_{ks}^+ |\{0\}\rangle \langle\{0\}| a_{ks} \rangle_t$$
$$= \langle a_{ks}^+ : \prod_{ks} e^{-a_{ks}^+ a_{ks}} : a_{ks} \rangle_t , \tag{10.10}$$

where two colons indicate the normal ordering and we expressed the vacuum state in terms of the creation and annihilation operators. Because of the presence of only one photon (10.10) reduces to

$$p_{ks}(t) = \langle a_{ks}^+(t) \, a_{ks}(t) \rangle$$
$$= |g_{ks}|^2 [(\omega_{ks}-\omega_0)^2 + \gamma^2]^{-1} \{1 + e^{-2\gamma t} - 2e^{-\gamma t}\cos(\omega_{ks}-\omega_0)t\} \tag{10.11}$$
$$\cdot \langle S^+(0) S^-(0) \rangle ,$$

$$\xrightarrow{\quad t\to\infty \quad} |g_{ks}|^2 \{(\omega_{ks}-\omega_0)^2 + \gamma^2\}^{-1} \langle S^+(0) S^-(0) \rangle . \tag{10.12}$$

Hence the spectrum is Lorentzian, centered at $\omega_{ks}=\omega_0$ and with the half-width equal to γ; this coincides with the result obtained by other theories (§ 3—5). The only difference is that we have now succeeded in obtaining the Lamb shift of the ground state as well, which was missing from earlier formulas, for example (3.12). Note also that, if Eq. (7.1) is used to calculate $N_{ks}(t)$ in the lowest nonvanishing order, then we have

$$N_{ks}(t) = |g_{ks}|^2 \int\limits_0^t d\tau \, e^{-2\gamma\tau}(1 - e^{-iX(t-\tau)}) \, (iX)^{-1} + \text{c.c.}$$
$$X = \omega_{ks} - \omega , \tag{10.13}$$

which is very different from (10.11); moreover, the Lamb shift is completely missing from (10.13). It thus appears that the Born approximation cannot be made on (7.1).

We now consider the far-field behavior. The normally ordered correlation function is given by (7.14):

$$\langle E^{(-)}(r, t + |r|/c) E^{(+)}(r', t' + |r'|/c) \rangle = k_0^4 |r|^{-1} |r'|^{-1}$$
$$\cdot [\hat{r} \times (\hat{r} \times d)] [\hat{r}' \times (\hat{r}' \times d)] \langle S^+(t) S^-(t') \rangle = \psi(r, t) \, \psi^*(r', t') . \tag{10.14a}$$

The first-order coherence function factorizes and hence the field possesses *first-oder coherence* and no higher orders due to the presence of only one photon. In particular, we have

$$\langle E^{(-)}(r, t + |r|/c) \cdot E^{(+)}(r, t + |r|/c) \rangle$$
$$= (r^{-2} k_0^4 |d|^2 \sin^2 \theta) \exp\{i\omega_0(t - t') - \gamma(t + t')\}, \quad (t > t'), \tag{10.14b}$$

where we assume that the atom at time $t = 0$ was in the excited state and θ is the angle between r and d. It is interesting that the correlation (10.14) is of the same form as that for a damped Hertzian dipole[17].

Finally, we make few comments on the Langevin equations. The Langevin equations (8.4) reduce to

$$\dot{S}^{\pm} = (\pm i\omega_0 - \gamma) S^{\pm} + F^{\pm}, \tag{10.15}$$

where

$$\langle F^{\pm}(t) \rangle = 0, \langle F^+(t) F^-(t') \rangle = 0, \quad \langle F^-(t) F^+(t') \rangle = 2\gamma\delta(t - t'). \tag{10.16}$$

The Langevin equations are linear and are solved easily

$$S^{\pm}(t) = e^{(\pm i\omega_0 - \gamma)t} S^{\pm}(0) + \int_0^t F^{\pm}(t - \tau) e^{(\pm i\omega_0 - \gamma)\tau} d\tau. \tag{10.17}$$

Hence the correlation function is given by

$$\langle S^+(t) S^-(t') \rangle = \exp\{i\omega_0(t - t') - \gamma(t + t')\}; \tag{10.18}$$

the other terms do not contribute because of (10.16) and $\langle S^+(0) F^-(t) \rangle = 0$, on the other hand the correlation function $\langle S^-(t) S^+(t') \rangle$ is

$$\langle S^-(t) S^+(t') \rangle = \int_0^t dt_1 \int_0^{t'} dt_2 \exp\{-(i\omega_0 + \gamma)(t - t_1) - (-i\omega_0 + \gamma)(t' - t_2)\}$$
$$\cdot 2\gamma\delta(t_1 - t_2) = e^{-i\omega_0(t - t')}[e^{-\gamma(t - t')} - e^{-\gamma(t + t')}], \tag{10.19}$$

where we used the initial condition $\langle S^-(0) S^+(0) \rangle = 0$. (10.18) and (10.19) are, of course, identical to the results obtained from the regression theorem. The above also shows that

$$\langle [S^+(t), S^-(t)] \rangle = 2 \langle S^z(t) \rangle, \tag{10.20}$$

i.e. the atomic commutation relations are valid[18]. The c-number Langevin equation is even simpler [cf. (8.19)]

$$\dot{z} = -(i\omega_0 + \gamma) z, \tag{10.21}$$

---

[17] The correlation function as given by (10.14) differs by a factor of two from the one obtained in [14].

[18] One should really show also that the equal-time commutation relations between the matter and field operators vanish for all time in the same limit in which the master equation is valid. This can be shown to be so by working out several commutators in Born and Markov approximations.

implying that

$$z(t) = e^{-(i\omega_0 + \gamma)(t-t')} z(t'), \tag{10.22}$$

which, of course, immediately leads to the normally ordered expectation values

$$\langle S^{\pm}(t) \rangle = e^{(\pm i\omega_0 - \gamma)(t-t')} \langle S^{\pm}(t') \rangle,$$

$$\langle S^+(t) S^-(t') \rangle = e^{i\omega_0(t-t') - \gamma(t+t')} \langle S^+(0) S^-(0) \rangle. \tag{10.23}$$

11. Spontaneous Emission from Two Two-Level Atoms

We now consider the spontaneous emission from a system of two atoms and discuss some of the dynamical aspects of the cooperative effects. The master equation (6.53) and the Langevin equations (8.19) reduce to (suppressing the subscript from ω_0)

$$\begin{aligned}
\partial\varrho/\partial t = & -i\omega[S_1^z + S_2^z, \varrho] - i\Omega_{12}[S_1^+ S_2^- + S_2^+ S_1^-, \varrho] \\
& - \gamma \sum_{i=1,2} (S_i^+ S_i^- \varrho - 2S_i^- \varrho S_i^+ + \varrho S_i^+ S_i^-) \\
& - \gamma_{12}\{(S_1^+ S_2^- + S_2^+ S_1^-)\varrho \\
& + \varrho(S_1^+ S_2^- + S_2^+ S_1^-) - 2S_1^- \varrho S_2^+ - 2S_2^- \varrho S_1^+\},
\end{aligned} \tag{11.1}$$

$$\dot{z}_1 = -(i\omega + \gamma)z_1 + (\gamma_{12} + i\Omega_{12})z_2(2|z_1|^2 - 1), \tag{11.2a}$$

$$\dot{z}_2 = -(i\omega + \gamma)z_2 + (\gamma_{12} + i\Omega_{12})z_1(2|z_2|^2 - 1). \tag{11.2b}$$

From (11.2) we obtain easily the equations of motion. Some of the equations needed to calculate the radiation rate are

$$\begin{aligned}
\langle \dot{S_1^+ S_1^-} \rangle + 2\gamma \langle S_1^+ S_1^- \rangle = & -\gamma_{12}\langle S_1^+ S_2^- + S_2^+ S_1^- \rangle \\
& - i\Omega_{12}\langle S_1^+ S_2^- - S_2^+ S_1^- \rangle,
\end{aligned} \tag{11.3a}$$

$$\begin{aligned}
\langle \dot{S_2^+ S_2^-} \rangle + 2\gamma \langle S_2^+ S_2^- \rangle = & -\gamma_{12}\langle S_1^+ S_2^- + S_2^+ S_1^- \rangle \\
& + i\Omega_{12}\langle S_1^+ S_2^- - S_2^+ S_1^- \rangle,
\end{aligned} \tag{11.3b}$$

$$\begin{aligned}
\langle \dot{S_1^+ S_2^-} \rangle + 2\gamma \langle S_1^+ S_2^- \rangle = & 4\gamma_{12}\langle S_1^+ S_1^- S_2^+ S_2^- \rangle \\
& - i\Omega_{12}\langle S_1^+ S_1^- - S_2^+ S_2^- \rangle \\
& - \gamma_{12}\langle S_1^+ S_1^- + S_2^+ S_2^- \rangle,
\end{aligned} \tag{11.4}$$

$$\langle S_1^+ S_1^- \dot{S_2^+ S_2^-} \rangle = -4\gamma \langle S_1^+ S_1^- S_2^+ S_2^- \rangle. \tag{11.5}$$

These equations form a closed set. From (11.3) and (11.4) we obtain immediately

$$\begin{pmatrix} \langle S_1^+ S_1^- \rangle - \langle S_2^+ S_2^- \rangle \\ \langle S_1^+ S_2^- \rangle - \langle S_2^+ S_1^- \rangle \end{pmatrix} = e^{-2\gamma t} \begin{pmatrix} \cos 2\Omega_{12}t & -i\sin 2\Omega_{12}t \\ -i\sin 2\Omega_{12}t & \cos 2\Omega_{12}t \end{pmatrix} \begin{pmatrix} \alpha_1 \\ \alpha_2 \end{pmatrix}, \qquad (11.6)$$

where α_1 and α_2 are determined from the initial condition. (11.5) leads to the simple decay for the probability that each atom remains in the excited state:

$$\langle S_1^+ S_1^- S_2^+ S_2^- \rangle = e^{-4\gamma t} \beta, \qquad (11.7)$$

and then we obtain from (11.3) and (11.4)

$$(\partial/\partial t) \langle S_1^+ S_1^- + S_2^+ S_2^- \rangle + 2\gamma \langle S_1^+ S_1^- + S_2^+ S_2^- \rangle \\ + 2\gamma_{12} \langle S_1^+ S_2^- + S_2^+ S_1^- \rangle = 0, \qquad (11.8a)$$

$$(\partial/\partial t) \langle S_1^+ S_2^- + S_2^+ S_1^- \rangle + 2\gamma \langle S_1^+ S_2^- + S_2^+ S_1^- \rangle \\ + 2\gamma_{12} \langle S_1^+ S_1^- + S_2^+ S_2^- \rangle = 8\gamma_{12} \beta e^{-4\gamma t}. \qquad (11.8b)$$

On solving (11.8) we find that

$$\lambda(t) = \langle S_1^+ S_1^- + S_2^+ S_2^- + S_1^+ S_2^- + S_2^+ S_1^- \rangle \\ = e^{-2(\gamma + \gamma_{12})t} \{ \lambda_0 + 4\gamma_{12} \beta (\gamma - \gamma_{12})^{-1} (1 - e^{-2(\gamma - \gamma_{12})t}) \}, \qquad (11.9a)$$

$$\langle S_1^+ S_1^- + S_2^+ S_2^- \rangle_t = e^{-2(\gamma - \gamma_{12})t} \langle S_1^+ S_1^- + S_2^+ S_2^- \rangle_0 \\ - 2\gamma_{12} \int_0^t e^{-2(\gamma - \gamma_{12})(t - \tau)} \lambda(\tau) \, d\tau. \qquad (11.9b)$$

We examine two special cases. First, we consider the case when one atom was in the excited state ($|1\rangle$) and the other in the ground state ($|2\rangle$), i.e.

$$\varrho(0) = |1, 2\rangle \langle 1, 2|, \qquad (11.10a)$$

then from (11.6) and (11.9) we have

$$\beta = 0, \quad \alpha_1 = 1, \quad \alpha_2 = 0, \quad \lambda_0 = 1, \quad \lambda(t) = e^{-2(\gamma + \gamma_{12})t},$$

$$\langle S_1^+ S_1^- \rangle - \langle S_2^+ S_2^- \rangle = e^{-2\gamma t} \cos 2\Omega_{12}t, \qquad (11.10b)$$

$$\langle S_1^+ S_1^- \rangle_t = \tfrac{1}{2} e^{-2\gamma t} \cos 2\Omega_{12}t + \tfrac{1}{4} \{ e^{-2(\gamma - \gamma_{12})t} + e^{-2(\gamma + \gamma_{12})t} \},$$

$$\langle S_2^+ S_2^- \rangle_t = -\tfrac{1}{2} e^{-2\gamma t} \cos 2\Omega_{12}t + \tfrac{1}{4} \{ e^{-2(\gamma - \gamma_{12})t} + e^{-2(\gamma + \gamma_{12})t} \}. \qquad (11.10c)$$

In the steady state

$$\langle S_1^z \rangle = \langle S_2^z \rangle = -\tfrac{1}{2} \quad \text{unless} \quad \gamma = \gamma_{12}, \qquad (11.10d)$$

implying that each atom is found in the ground state unless $\gamma = \gamma_{12}$. The limit $\gamma = \gamma_{12}$ is strictly never reached unless two atoms are at the same site in which case

$$\langle S_1^z \rangle = \langle S_2^z \rangle = 0 . \tag{11.10e}$$

The radiation rate defined by (7.18) is in the present case

$$I(t) = 2\omega \{ (\gamma - \gamma_{12}) e^{-2(\gamma - \gamma_{12})t} + (\gamma + \gamma_{12}) e^{-2(\gamma + \gamma_{12})t} \} . \tag{11.10f}$$

We next treat the case when each of the atoms was prepared in its excited state, i.e.

$$\varrho(0) = |1, 1\rangle \langle 1, 1| \Rightarrow \alpha_1 = 0, \quad \alpha_2 = 0, \quad \beta_0 = 1, \quad \lambda_0 = 2, \tag{11.11a}$$

then

$$\lambda(t) = 2 e^{-2(\gamma + \gamma_{12})t} + 4\gamma_{12}(\gamma - \gamma_{12})^{-1} [e^{-2(\gamma + \gamma_{12})t} - e^{-4\gamma t}], \tag{11.11b}$$

$$\langle S_1^+ S_1^- \rangle_t = \langle S_2^+ S_2^- \rangle_t = e^{-2(\gamma - \gamma_{12})t} - \tfrac{1}{2}(\gamma + \gamma_{12})(\gamma - \gamma_{12})^{-1} [e^{-2(\gamma - \gamma_{12})t}$$
$$- e^{-2(\gamma + \gamma_{12})t}] + 2\gamma_{12}\gamma_{12}(\gamma^2 - \gamma_{12}^2)[e^{-2(\gamma - \gamma_{12})t} - e^{-4\gamma t}], \tag{11.11c}$$

which implies that at each time there is equipartition of the energy and each is left in its ground state at $t = \infty$ unless we have the limiting case $\gamma = \gamma_{12}$. This limiting case appears to be somewhat unphysical. It is interesting to note that the radiation rate $I(t)$ and the energy or correlations like $\langle S_1^+ S_2^- \rangle$ are independent of the cooperative shift Ω_{12}, whereas in the case of excitation (11.10a) the energy of the individual atom or the correlation $\langle S_1^+ S_2^- \rangle$ (off-diagonal element) is dependent on Ω_{12}.

It should be noted that the Eq. (6.53) has a (rather trivial) symmetry property in that the Liouville operator remains invariant under the permutation of the particles if the positional coordinates are also permuted. This implies that, if initially

$$\varrho(\{S_i^+, S_i^-, S_i^z, r_i\}, 0) = \varrho(\Pi\{S_i^+, S_i^-, S_i^z, r_i\}, 0),$$

then for all times

$$\varrho(\{S_i^+, S_i^-, S_i^z, r_i\}, t) = \varrho(\Pi\{S_i^+, S_i^-, S_i^z, r_i\}, t),$$

where Π denotes the permutation of the indices $1, 2, ..., N$. For the two-atom problem where each atom is initially excited the above symmetry leads to $\langle S_1^+ S_1^- \rangle = \langle S_2^+ S_2^- \rangle$, $\langle S_1^+ \rangle = \langle S_2^+ \rangle$, $\langle S_1^+ S_2^- \rangle = \langle S_2^+ S_1^- \rangle$ etc., since the only parameters which enter the problem are γ_{ij}, Ω_{ij} and γ and these are symmetric under the interchange of 1 and 2.

A detailed discussion of the properties of the spontaneous emission from two two-level atoms can be found in the papers by Lehmberg [72], Stephen [43] and others [73]. We close this section by presenting the time dependence of the dipole moment. From (11.2) we obtain easily the relevant equations of motion

$$\langle \dot{S_1^- \pm S_2^-} \rangle = -\{(i\omega + \gamma) \pm (\gamma_{12} + i\Omega_{12})\} \langle S_1^- \pm S_2^- \rangle + 2(\gamma_{12} + i\Omega_{12})$$
$$\cdot \langle S_1^+ S_1^- S_2^- \pm S_2^+ S_2^- S_1^- \rangle, \tag{11.12}$$

$$\langle \dot{S_2^+ S_2^- S_1^- \pm S_1^+ S_1^- S_2^-} \rangle = -\{(i\omega + 3\gamma) \pm (\gamma_{12} - i\Omega_{12})\}$$
$$\cdot \langle S_2^+ S_2^- S_1^- \pm S_1^+ S_1^- S_2^- \rangle. \tag{11.13}$$

On solving (11.12), (11.13) we find that the time dependence of the dipole moment is given by

$$\langle S_1^- \pm S_2^- \rangle_t = \exp\{-[i\omega + \gamma \pm (\gamma_{12} + i\Omega_{12})]t\} [\langle S_1^- \pm S_2^- \rangle_0$$
$$\pm (\gamma_{12} + i\Omega_{12})(\gamma \mp i\Omega_{12})^{-1} (1 - \exp\{-(\gamma \mp i\Omega_{12})t\})$$
$$\cdot \langle S_1^- S_2^+ S_2^- \pm S_2^- S_1^+ S_1^- \rangle. \tag{11.14}$$

It is easily checked that for each of the excitations (11.10a) and (11.11a) the dipole moment of each atom remains zero. The two-time correlation functions like $\langle S_i^+(t) S_j^-(t') \rangle$ can be obtained from (11.14) and the quantum regression theorem. One has, for instance

$$\langle S_1^+(\tau) S_1^-(t) \rangle = \tfrac{1}{2}\exp\{-(i\omega + \gamma + \gamma_{12} + i\Omega_{12})(t - \tau)\} [\langle S_1^+ S_1^- + S_1^+ S_2^- \rangle_\tau$$
$$+ (\gamma_{12} + i\Omega_{12})(\gamma - i\Omega_{12})^{-1} (1 - \exp\{-(\gamma - i\Omega_{12})(t - \tau)\})$$
$$\cdot \langle S_1^+ S_1^- S_2^+ S_2^- \rangle_\tau]$$
$$+ \tfrac{1}{2}\exp\{-(i\omega + \gamma - \gamma_{12} - i\Omega_{12})(t - \tau)\}$$
$$\cdot [\langle S_1^+ S_1^- - S_1^+ S_2^- \rangle_\tau$$
$$- (\gamma_{12} + i\Omega_{12})(\gamma + i\Omega_{12})^{-1} (1 - \exp\{-(\gamma + i\Omega_{12})(t - \tau)\})$$
$$\cdot \langle S_1^+ S_1^- S_2^+ S_2^- \rangle_\tau], \quad (\tau < t), \tag{11.15}$$

where the one-time mean values appearing in (11.15) are obtained from (11.6), (11.7) and (11.9).

12. Emission from a System of Harmonic Oscillators

We have already presented the master equation and the Langevin equations describing spontaneous emission from a system of harmonic oscillators. The linearity of the Langevin equations makes this model rather attractive and a number of features of cooperative phenomena can be studied in some detail. Some of the cooperative effects in a system of two harmonic oscillators have been observed by Lama et al. [74]. Moreover, the present model offers us a deep insight into the problem of emission from a system of two-level atoms.

We found in Chapter 8 that the Sudarshan-Glauber distribution function Φ satisfies the equation

$$\partial\Phi/\partial t = (i\omega_0 + \gamma)\sum_i (\partial/\partial z_i)(z_i\Phi) + \sum_{i\neq j}(\gamma_{ij} + i\Omega_{ij})(\partial/\partial z_i)(z_j\Phi) + \text{c.c.} \quad (12.1)$$

The corresponding Langevin equations are given by (8.27), viz.

$$\dot{z}_i = -(i\omega_0 + \gamma)z_i - \sum_{j\neq i}(\gamma_{ij} + i\Omega_{ij})z_j. \quad (12.2)$$

The solution of (12.2) is given by

$$z(t) = \exp\{-(i\omega_0 + \gamma)t\}\exp\{-\Gamma t\}z(0), \quad (12.3)$$

where

$$\Gamma_{ij} = (1 - \delta_{ij})(\gamma_{ij} + i\Omega_{ij}). \quad (12.4)$$

The solution of (12.1) subject to the initial condition

$$\Phi(\{z\}, \{z^*\}, 0) = \prod_i \delta^{(2)}(z_i - z_i^0), \quad (12.5)$$

is given by

$$K(\{z_i\}, \{z_i^*\}, t \mid \{z_i^0\}, \{z_i^0{}^*\}, 0) = \prod_i \delta^{(2)}(z_i - z_i(t)), \quad (12.6)$$

where

$$z(t) = \exp\{-(i\omega_0 + \gamma)t\}\exp\{-\Gamma t\}z^0, \quad (12.7)$$

and where we have denoted the solution of (12.1) subject to (12.5) by K, which is simply the Green's function corresponding to (12.1). The knowledge of K is sufficient to permit us to calculate all the statistical properties of the oscillator system. The state of the system at time t is given by

$$\Phi(\{z_i\}, \{z_i^*\}, t) = \int d^2\{z_i^0\}\, K(\{z_i\}\{z_i^*\}, t \mid \{z_i^0\}, \{z_i^0{}^*\}, 0)\, \Phi_0(\{z_i^0\}, \{z_i^0{}^*\}),$$

where Φ_0 represents the initial state.

We first calculate the radiation rate, which is given by (7.17) or by (7.18). For the present model it is

$$I(t) = 2\omega\sum_{ij}\gamma_{ij}\langle a_i^+(t)\, a_j(t)\rangle, \quad (12.8)$$

which, on using the solution of the Langevin equation, becomes

$$I(t) = 2\omega e^{-2\gamma t}\sum_{ij}(e^{-\Gamma^+ t}\gamma e^{-\Gamma t})_{ij}\langle a_i^+(0)\, a_j(0)\rangle, \quad (12.9)$$

where the matrix γ has elements γ_{ij}. We rewrite (12.9) as

$$I(t) = I_1 + I_2 , \tag{12.10}$$

$$I_1 = 2\omega e^{-2\gamma t} \sum_i (e^{-\Gamma^+ t} \gamma e^{-\Gamma t})_{ii} \langle a_i^+(0) a_i(0) \rangle , \tag{12.11}$$

$$I_2 = 2\omega e^{-2\gamma t} \sum_{i \neq j} (e^{-\Gamma^+ t} \gamma e^{-\Gamma t})_{ij} \langle a_i^+(0) a_j(0) \rangle . \tag{12.12}$$

The first term I_1 is evidently the incoherent part of the radiation rate and the second term is the coherent part. The coherent part will be non-zero only if the initial state is such that

$$\langle a_i^+(0) a_j(0) \rangle \neq 0 \quad \text{for} \quad i \neq j . \tag{12.13}$$

If the initial state is the one in which there are no correlations, i.e.

$$\langle a_i^+(0) a_j(0) \rangle = \langle a_i^+(0) \rangle \langle a_j(0) \rangle ,$$

then (12.13) holds only if the initial dipole moment is not zero; this situation has been referred to as superradiance of the first kind (cf. Chapter 9). If the dipole moment is zero, then the initial state must be one with correlations, otherwise $I_2 = 0$ (superradiance of the second kind). The question we now ask is: What are the superradiant states of the harmonic oscillator system? To discuss these, we examine the perturbative results

$$I_0 = 2\omega\gamma \sum_{ij} \langle a_i^+(0) a_j(0) \rangle , \tag{12.14}$$

for the case of a system confined to a region smaller than a wavelength. We can rewrite (12.14) as

$$I_0 = 2\omega\gamma \langle D^+(0) D(0) \rangle , \tag{12.15}$$

where we introduced the collective operator D defined by

$$D = \sum_i a_i . \tag{12.16a}$$

Note that this operator satisfies the harmonic oscillator commutation relations (except for normalization)

$$[D, D^+] = N, [D, D] = [D^+, D^+] = 0 , \tag{12.16b}$$

and so one can also introduce the "normalized" operator

$$\mathscr{D} = D/\sqrt{N} = (1/\sqrt{N}) \Sigma a_i . \tag{12.16c}$$

The operators D and D^+ are the analog of Dicke's collective operators S^- and S^+ (8). We can also introduce the coherent states and the Fock states associated with $\mathscr{D}$. Consider the state defined by [75]

$$\Psi_m = \sum_{\{m\}} \delta_{m.\Sigma m_i} c_{\{m\}} e^{i\varphi(\{m\})} |\{m\}\rangle . \tag{12.17a}$$

It is evident that

$$\sum_i a_i^+ a_i \Psi_m = m \Psi_m , \tag{12.18}$$

$$\sum_{j \neq l} a_j^+ a_l \Psi_m = \sum_{\{m\}} \delta_{m.\Sigma m_i} \sum_{j \neq l} c(\{m_j - 1, m_l + 1\})$$

$$\cdot \sqrt{m_j(m_l + 1)} \, e^{i\varphi(\{m_j - 1.m_l + 1\})} |\{m\}\rangle . \tag{12.19}$$

If we choose

$$\varphi(\{m\}) = \varphi_0 , \qquad c_{\{m\}} = \frac{m!}{(m_1! m_2! \ldots m_N!)^{\frac{1}{2}}} \mathcal{N}_m , \tag{12.17b}$$

where $\mathcal{N}_m$ is the normalization factor, then

$$\sum_{j \neq l} a_j^+ a_l \Psi_m = (N - 1) m \Psi_m , \tag{12.17c}$$

and hence

$$D^+ D \Psi_m = N m \Psi_m . \tag{12.19}$$

Hence Ψ_m is an eigenstate of $D^+ D$ with the eigenvalue Nm. For the initial state defined by (12.17), the radiation rate is

$$I_0 = 2\omega\gamma N m . \tag{12.20}$$

It should be noted that $m\omega$ is the total initial energy of the atomic system. (12.20) indeed shows the superradiant behavior. If the phases $\varphi(\{m\})$ are completely random, implying an initial density operator of the form

$$\varrho_m = \sum_{\{m\}} \delta_{m.\Sigma m_i} c_{\{m\}} c_{\{m\}}^* |\{m\}\rangle \langle\{m\}| , \tag{12.21}$$

then the emission is normal. In the state Ψ_m the dipole moment is zero.

Next, if the system is excited initially to a coherent state, as is the case with excitation by external fields,

$$\Psi_{coh} = |z_1 z_2 \ldots z_N\rangle , \tag{12.22}$$

then, restricting ourselves for the sake of simplicity to the case of small samples, the initial radiation rate is

$$I_0 = 2\omega\gamma |\Sigma z_i|^2 \tag{12.23}$$

hence, if each of the oscillators is similarly excited $z_i = z_0$, then the radiation is superradiant. The coherent-state excitation is, of course, the one in which the dipole moment is non-zero. Again if the amplitude of excitation z_i had a random phase, then the emission would be incoherent. It is certainly much easier to prepare the system in the state (12.22) than in the state (12.17a) with the coefficients given by (12.17b).

The dynamical aspects of the radiation rate are given by (12.9) which shows simple decay behavior. For the case of two oscillators it can be shown that

$$\left[\begin{pmatrix} \gamma & \gamma_{12} \\ \gamma_{12} & \gamma \end{pmatrix}, \begin{pmatrix} 0 & \Omega_{12} \\ \Omega_{12} & 0 \end{pmatrix} \right] = 0 . \tag{12.24}$$

The radiation rate I_F in the case when the system was initially excited to a state with "no" correlations and zero dipole moment is

$$I_F(t) = 2\omega \{\gamma \cosh(2\gamma_{12}t) - \gamma_{12} \sinh(2\gamma_{12}t)\} e^{-2\gamma t} \sum_i \langle a_i^+(0) a_i(0) \rangle , \tag{12.25}$$

whereas for the case (12.22) with $|z_1| = |z_2|$ and a relative phase difference φ_0, the radiation rate I_{coh} is

$$I_{\text{coh}} = 2\omega |z_0|^2 e^{-2\gamma t} \{(\gamma - \gamma_{12}) e^{2\gamma_{12}t} (1 - \cos\varphi_0)$$
$$+ (\gamma + \gamma_{12}) e^{-2\gamma_{12}t} (1 + \cos\varphi_0)\} . \tag{12.26}$$

These radiation rates do not depend on the cooperative frequency shift Ω_{12}, which is due to (12.24). Thus Eqs. (12.9) and (12.3) show that the calculation of the properties of the harmonic oscillator model is essentially a problem in the theory of matrices. For the case of small samples where the effect of Ω_{ij} is ignored, it is simple to calculate $e^{-\Gamma t}$ because now

$$\Gamma_{ij} = \gamma(1 - \delta_{ij}) , \tag{12.27a}$$

and it is easy to show that

$$(e^{-\Gamma t})_{ij} = e^{\gamma t} \{\delta_{ij} - N^{-1}(1 - e^{-N\gamma t})\} . \tag{12.27b}$$

The Green's function K now reduces to

$$K(\{z_i\}, \{z_i^*\}, t | \{z_i^0\}, \{z_i^{0*}\}, 0) = \prod_i \delta^{(2)}(z_i - z_i(t)) , \tag{12.28}$$

$$z_i(t) = e^{-i\omega_0 t} \{z_i^0 - N^{-1}(1 - e^{-N\gamma t}) \sum z_j^0\} .$$

It should be noted that (12.28) is also the solution of the Langevin equation. Moreover, (12.28) shows that in the steady state

$$\varrho(\infty) = \prod_i |z_i^0 - N^{-1} \sum z_j^0\rangle_{i\, i}\langle z_i^0 - N^{-1} \sum z_j^0| , \tag{12.29a}$$

i.e. each oscillator is not left in its vacuum state (ground state), which is so only if $z_j^0 = z_0$ for all j. This is as expected since in the present case the coupling in the Hamiltonian is via the collective operator D. In view of this, one would expect the system to be left in the ground state of D, which is indeed the case (cf. Appendix C).

$$D\varrho(\infty) = 0. \tag{12.29b}$$

We now consider some special cases.

A) Initial Excitation Given by $\varrho(0) = |\{z_0\}\rangle \langle\{z_0\}|$

We first consider the case when each oscillator has been initially excited (similarly) to the coherent state $|z_0\rangle$. The density operator at time t is given by

$$\varrho(t) = \prod_i |z_0\, e^{-N\gamma t - i\omega_0 t}\rangle_i \,_i\langle z_0\, e^{-N\gamma t - i\omega_0 t}|, \tag{12.30}$$

showing that the system remains in a coherent state with the amplitude of oscillation decaying (the decay constant is N times that of a single oscillator). There are no correlations induced among any two oscillators. The radiation rate is given by

$$I(t) = 2\omega\gamma N^2\, e^{-2N\gamma t} |z_0|^2, \tag{12.31}$$

which is proportional to the square of the number of oscillators. The two-time correlation function for the operator $D(t)$ is given by

$$\langle D^+(t)\, D(t')\rangle = e^{i\omega_0(t-t') - N\gamma(t+t')} N^2 |z_0|^2, \tag{12.32}$$

where in obtaining (12.32) we used the solution (12.3) specialized to the present excitation. One can similarly show that

$$\langle D^+(t)\, D^+(t)\, D(t')\, D(t')\rangle = e^{2i\omega_0(t-t') - 2N\gamma(t+t')} N^4 |z_0|^4. \tag{12.33}$$

B) Initial Excitation Given by the Fock State ψ_m

In this case one finds on using (12.28) and (12.17) that

$$\langle D(t)\rangle = \langle D^+(t)\rangle = 0,$$

$$\langle D^+(t) D(t)\rangle = N m e^{-2N\gamma t}, \quad I(t) = 2\omega\gamma N m e^{-2N\gamma t},$$

$$\langle D^+(t) D(t')\rangle = e^{-N\gamma(t+t') + i\omega_0(t-t')} N m, \tag{12.34}$$

$$\langle D^+(t) D^+(t) D(t) D(t)\rangle = e^{-4N\gamma t} N m(N m - 1),$$

which show the time-dependent properties. On the other hand, for the case of pure Fock state, the excitation $c_{(m)} = 0$ except for one set of values, i.e.

$$\Psi_F = |m_1 m_2 \ldots m_N\rangle$$

and one finds for the radiation rate

$$I_F(t) = 2\gamma\omega e^{-2N\gamma t} \Sigma m_i, \tag{12.35}$$

the normal incoherent value. Moreover, one finds that, due to spontaneous emission, correlations are induced among any two oscillators, e.g.

$$\langle a_i^+(t) a_j(0)\rangle - \langle a_i^+(t)\rangle \langle a_j(0)\rangle = -(m_j/N)(1 - e^{-N\gamma t}) \quad (i \neq j). \tag{12.36}$$

We also discuss some of the normally ordered correlation functions for the electric field in the radiation zone. We have from (7.13), specialized to the case of small samples,

$$E^{(+)}(r, t) \sim E_0^{(+)}(r, t) - k_0^2(\hat{r} \times (\hat{r} \times d)/r) D(t - |r|/c), \tag{12.37}$$

and hence

$$\langle E^{(-)}(r, t) E^{(+)}(r', t')\rangle = (k_0^4/rr')(\hat{r} \times (\hat{r} \times d))(\hat{r}' \times (\hat{r}' \times d))$$

$$\cdot \langle D^+(t - |r|/c) D(t' - |r'|/c)\rangle. \tag{12.38}$$

For the coherent-state excitation case A, (12.38) reduces to

$$\langle E^{(-)}(r, t) E^{(+)}(r', t')\rangle = (k_0^4/rr')(\hat{r} \times (\hat{r} \times d))(\hat{r}' \times (\hat{r}' \times d))$$

$$\cdot N^2|z_0|^2 \exp\{i\omega_0(t - |r|/c - t' + |r'|/c) - N\gamma(t + t' - |r|/c - |r'|/c)\}, \tag{12.39}$$

which is typical of the damped Hertzian dipole. N^2 dependence is to be noted. Higher-order correlation functions can be computed by means of (12.33). It is easily seen that the spontaneously emitted field is a "coherent" field. The photon-counting distribution of such a field is Poissonian [30]. For excitation case B, the normally ordered correlation function is still given by (12.39) with $N|z_0|^2$ replaced by m. The fourth-order correlation function in this case will in view of (12.34) involve the factor $Nm(Nm - 1)$ which is, of course, characteristic of a field with a *fixed number of photons equal to Nm*. This is to be expected since in case B the atomic excitation has a finite value, whereas in case A the atomic excitation has an indefinite value. Other features of spontaneous emission from such a system can be found in Ref. [30].

Finally we discuss briefly the resolution of the harmonic oscillator paradox (cf. Chapter 3). The mean number of photons in the mode ks

for the case of a single oscillator is given by (7.23), viz.

$$N_{ks}(t) = |g_{ks}|^2 \int_0^t \int_0^{t_1} dt_1 dt_2 e^{-i\omega_{ks}(t_1-t_2)} \langle a^+(t_1) a(t_2) \rangle + \text{H.C.}$$

This with the use of the solution $z(t) = e^{-i\omega_0 t - \gamma t} z(0)$ of the Langevin equation for a single oscillator reduces to

$$N_{ks}(t) = |g_{ks}|^2 [(\omega_{ks} - \omega_0)^2 + \gamma^2]^{-1} \{1 + e^{-2\gamma t} - 2e^{-\gamma t} \cos(\omega_{ks} - \omega_0)t\}$$

$$\cdot \langle a^+(0) a(0) \rangle, \tag{12.40}$$

where ω_0 is the renormalized frequency given by (6.66). Equation (12.40) shows that the linewidth is equal to γ and thus resolves the apparent paradox.

13. Emission from a Small Sample of Two-Level Atoms; Master Equation: Exact Solution

We consider in the present section and in the next the dynamical aspects of the cooperative emission from a small sample of two-level atoms. This is also the system which was studied by Dicke in the framework of perturbation theory. For small samples the master equation (6.53) reduces [cf. (6.56)] to

$$\partial \varrho / \partial t = -i\omega_0 [S^z, \varrho] - i \sum_{i \neq j} V_{ij} [S_i^+ S_j^-, \varrho]$$

$$- \gamma (S^+ S^- \varrho - 2S^- \varrho S^+ + \varrho S^+ S^-), \tag{13.1}$$

where we introduced the collective operators $S^\pm$, S^z. We see that for small samples the dipole-dipole interaction (static part) still makes a contribution. It is evident from (13.1) that

$$(\partial/\partial t) \langle S^2 \rangle = i \sum_{i \neq j} V_{ij} \langle [S_i^+ S_j^-, S^2] \rangle,$$

and since

$$\langle S^2 \rangle = \tfrac{3}{4} N + \sum_{i \neq j} \langle S_i \cdot S_j \rangle = \tfrac{3}{4} N + \sum_{i \neq j} \langle S_i^z S_j^z = \tfrac{1}{2} S_i^+ S_j^- + \tfrac{1}{2} S_j^+ S_i^- \rangle, \tag{13.2}$$

one can easily check that

$$(\partial/\partial t) \langle S^2 \rangle = 2i \sum_{i \neq j \neq l} V_{ij} \langle S_j^z \{S_i^+ S_l^- - S_l^+ S_i^- \} \rangle. \tag{13.3}$$

This relation shows that S^2 is not a constant of motion, provided that we are dealing with a system of only *two* atoms. In the case originally discussed by Dicke the variation of the phase factors from (2.12) was ignored so that (2.12) reduces to

$$H = \omega S^z + \sum_{ks} \omega_{ks} a_{ks}^+ a_{ks} + \sum_{ks} \{g_{ks} a_{ks}(S^+ + S^-) + \text{H.C.}\}. \tag{13.4}$$

It is clear from (13.4) that S^2 is a constant of motion. The master equation for the Dicke Hamiltonian in the Born, Markov and rotating-wave approximations is

$$\partial \varrho/\partial t + i\omega [S^z, \varrho] + \gamma(S^+ S^- \varrho - 2S^- \varrho S^+ + \varrho S^+ S^-) + i\Delta = 0, \tag{13.5}$$

where

$$\Delta = -\alpha_+ [S^+ S^-, \varrho] - \alpha_- [S^- S^+, \varrho], \tag{13.6a}$$

$$\alpha_\pm = \sum_{ks} |g_{ks}|^2 (\omega_{ks} \mp \omega)^{-1}. \tag{13.6b}$$

$\alpha_\pm$ are the frequency shift terms and of course to be renormalized. It should be noted that for the completely excited state $|\frac{1}{2}N, \frac{1}{2}N\rangle$ (ground state $|\frac{1}{2}N, -\frac{1}{2}N\rangle$) only α_+ (α_-) contributes. Arecchi and Kim [76] renormalize $\alpha_\pm$ by replacing them by

$$\alpha_\pm \approx \gamma k_c/\pi k_0 \tag{13.6c}$$

where k_c is the cutoff wave number, which is taken to be $2\pi/l_c$, with $l_c(\ll \lambda)$ standing for the size of the cooperative region. The master equation (13.5) has also been obtained by Bonifacio, Schwendimann and Haake, who adopted a different model for emission from needle-shaped samples. The values of the parameters are, of course, different in their model. The master equation (13.5) has been the subject of great many investigations [29, 31 to 33, 77 to 81]. We first present an exact solution of (13.5) and discuss its consequences.

On taking the matrix elements of (13.5) between Dicke states we obtain

$$\partial \varrho_{mn}/\partial \tau = -(i/2\gamma) [(m-n)\omega - \alpha_+(v_m - v_n) - \alpha_-(v_{m+1} - v_{n+1})] \varrho_{m.n} \tag{13.7}$$
$$+ (v_{m+1} v_{n+1})^{\frac{1}{2}} \varrho_{m+1.n+1} - \tfrac{1}{2}(v_m + v_n) \varrho_{m.n},$$

where

$$\varrho_{m.n} = \langle Sm|\varrho|Sn\rangle, \quad v_m = (S+m)(S-m+1), \quad \tau = 2\gamma t. \tag{13.8}$$

It should be noted that $2\gamma v_m$ is the transition probability per unit time (obtained by using Fermi's Golden Rule) that the system makes a transition from the state $|S, m\rangle$ to $|S, m-1\rangle$. For diagonal elements we have the Pauli type of equation

$$\partial \varrho_{mm}/\partial \tau = v_{m+1} \varrho_{m+1.m+1} - v_m \varrho_{m.m}. \tag{13.9}$$

A rate equation of the type (13.9) could have been obtained directly. However, we need the full master equation (13.7) to calculate the behavior of the dipole moment and the two-time correlation functions of the form $\langle S^+(t) S^-(t')\rangle$.

Equation (13.7) is in the form of a simple-difference differential equation[19] and can easily be solved by Laplace transformation and by iteration. Let $\hat{\varrho}_{m,n}(z)$ be the Laplace transform of $\varrho_{mn}(\tau)$ i.e.

$$\hat{\varrho}_{m,n}(z) = \int_0^\infty d\tau\, e^{-z\tau} \varrho_{m,n}(\tau), \quad \text{Re}\, z \geq 0, \tag{13.10}$$

then we have from (13.7)

$$\{z + \tfrac{1}{2}(v_m \beta_+ + v_n \beta_-) + i\omega'(m-n)\}\,\hat{\varrho}_{m,n} - \varrho_{m,n}(0) \tag{13.11a}$$

$$= (v_{m+1} v_{n+1})^{\frac{1}{2}} \hat{\varrho}_{m+1,n+1},$$

where

$$\beta_\pm = [\gamma \mp i(\alpha_- + \alpha_+)]/\gamma, \quad \omega' = (\omega + 2\alpha_-)/2\gamma. \tag{13.11b}$$

If we make a transformation to a representation in which $p = m - n$, $q = \tfrac{1}{2}(m + n)$, then we see that $\varrho_{p,q}$ is coupled to $\varrho_{p,q+1}$ only i.e. p remains fixed. Moreover, since by definition $\varrho_{\frac{1}{2}N+1,n} = \varrho_{m,\frac{1}{2}N+1} = 0$, it is clear that a general solution of (13.11a) will be of the form

$$\hat{\varrho}_{m,n}(z) = \sum_{l \geq 0} f_{m,n,l}(z)\, \varrho_{m+l,n+l}(0). \tag{13.12}$$

Then a straightforward analysis shows that

$$\hat{\varrho}_{m,n}(z) = \sum_{l \geq 0} \left(\prod_1^l v_{m+k} v_{n+k} \right)^{\frac{1}{2}} \prod_0^l [z + i\omega'(m-n)$$

$$+ \tfrac{1}{2}(\beta_+ v_{m+k} + v_{n+k}\beta_-)]^{-1} \varrho_{m+l,n+l}(0), \tag{13.13}$$

which is the exact solution of the master equation. For the diagonal elements (13.13) gives

$$\hat{\varrho}_{m,m}(z) = \sum_{l \geq 0} \left(\prod_1^l v_{m+k} \right) \prod_0^l [z + v_{m+k}]^{-1} \varrho_{m+l,m+l}(0). \tag{13.14}$$

It is evident from (13.13) that the poles of $\hat{\varrho}_{mn}(z)\,(m \neq n)$ are all *simple* poles whereas those of $\hat{\varrho}_{mm}(z)$ are of both first and second order. Inversion of (13.13) for the off-diagonal elements is rather simple, whereas the inversion of (13.14) is complicated by the presence of poles of order two.

---

[19] The master equation (13.9) is also basic in the treatment of simple "epidemics" and has been studied extensively by statisticians, see e.g. Bailey [82].

It is clear from (13.13) that if the density operator is diagonal initially, it remains diagonal for all times, i.e.

$$\varrho_{m.n}(0) \propto \delta_{mn} \Rightarrow \varrho_{m.n}(t) \propto \delta_{m.n}, \tag{13.15a}$$

and in particular the dipole moment remains zero for all times if it is zero initially

$$\langle S^\pm(0) \rangle = 0 \Rightarrow \langle S^\pm(t) \rangle = 0. \tag{13.15b}$$

The superradiant emission in this case would be due to the presence of atomic correlations [cf. our discussion following (9.6)]. The steady-state solution of (13.1) is

$$\varrho_{m.n}(\infty) = \delta_{mn} \delta_{m,-s}, \tag{13.16}$$

which can be understood easily if we recall that in the Hamiltonian the coupling through the radiation field is via the collective operators $S^\pm$ and we expect the system to decay to the ground state of S^z. The mean value of a normally ordered correlation of the form $\langle (S^+)^p (S^z)^q (S^-)^p \rangle$ is given by

$$\langle (S^+)^p (\hat{S}^z)^q (S^-)^p \rangle = \sum_m \left(\prod_0^p v_{m-k} \right) (m-p)^q \hat{\varrho}_{mm}. \tag{13.17}$$

For the initial excited state $\varrho(0) = |\frac{1}{2}N, \frac{1}{2}N\rangle \langle \frac{1}{2}N, \frac{1}{2}N|$ the Laplace transform of the energy is given by

$$\langle \hat{S}^z \rangle = \sum_{-\frac{1}{2}N}^{\frac{1}{2}N} m \frac{N!(\frac{1}{2}N-m)!}{(\frac{1}{2}N+m)!} \prod_{k=m}^{\frac{1}{2}N} (z+v_k)^{-1}, \tag{13.18}$$

and

$$\hat{\varrho}_{mm} = \frac{N!(\frac{1}{2}N-m)!}{(\frac{1}{2}N+m)!} \prod_{k=m}^{\frac{1}{2}N} (z+v_k)^{-1}. \tag{13.19}$$

In particular, for the three-atom problem one has $v_{\frac{3}{2}} = 3$, $v_{\frac{1}{2}} = 4$, $v_{-\frac{1}{2}} = 3$, $v_{-\frac{3}{2}} = 0$, and then

$$\varrho_{\frac{3}{2},\frac{3}{2}}(\tau) = e^{-3\tau}, \varrho_{\frac{1}{2},\frac{1}{2}}(\tau) = 3\{e^{-3\tau} - e^{-4\tau}\}, \varrho_{-\frac{1}{2},-\frac{1}{2}}(\tau) = 12\{e^{-4\tau}$$
$$+ \tau e^{-3\tau} - e^{-3\tau}\}, \varrho_{-\frac{3}{2},-\frac{3}{2}}(\tau) = (1 - 9e^{-4\tau} - 12\tau e^{-3\tau} + 8e^{-3\tau}), \tag{13.20}$$
$$\langle S^z(\tau) \rangle = -3e^{-3\tau} + 6e^{-4\tau} - \frac{3}{2} + 12\tau e^{-3\tau}.$$

The radiation rate $I(t)$ is given by the time rate change [20] of the energy [cf. (7.18)]

$$I(t) = -\omega(\partial/\partial t)\langle S^z \rangle = 2\gamma\omega\langle S^+ S^- \rangle,$$

[20] It is easily shown that, in the special cases where $N = 1, 2, 4,$ and 8, the solution (13.19) leads to the results of Dillard and Robl [83] (see also Dialetis [84]).

and its Laplace transform is given by

$$\hat{I} = \omega\{\langle S^z(0)\rangle - z\langle \hat{S}^z\rangle\}, \tag{13.21}$$

and therefore the steady-state value of the intensity is zero, as one would expect.

We next present the explicit form of the two-time correlation function $\langle S^+(t) S^-(t')\rangle$ $(t \geq t')$. From the regression theorem it is clear that [cf. Eq. (6.60)]

$$\langle S^+(t_1) S^-(t_2)\rangle = \sum_\alpha f_\alpha(t_1 - t_2) g_\alpha(t_2), \qquad (t_1 > t_2), \tag{13.22}$$

where f_α and g_α are the numerical functions. From (13.13) we can write

$$\langle S^+(\tau_1)\rangle = L_{\tau_1}^{-1} \mathrm{Tr}\{\hat{\varrho} S^+\} = L_{\tau_1}^{-1} \sum_m \hat{\varrho}_{m,m+1}(v_{m+1})^{\frac{1}{2}}$$

$$= \left\{ L_{\tau_1 - \tau_2}^{-1} \sum_m (v_{m+1})^{\frac{1}{2}} \sum_{l \geq 0} \left(\prod_1^l v_{m+k} v_{m+1+k}\right)^{\frac{1}{2}} \prod_0^l [z + \tfrac{1}{2}(\beta_+ v_{m+k} \right. \tag{13.23}$$

$$\left. + \beta_- v_{m+k+1}) - i\omega']^{-1}(v_{m+l+1})^{-\frac{1}{2}} \right\} \langle S^+|l+m\rangle \langle l+m|\rangle_{\tau_2},$$

where $L_\tau^{-1} f(z)$ denotes the inverse Laplace transform of

$$f(z) = L_\tau f(\tau) = \int_0^\infty f(\tau) e^{-z\tau} d\tau. \tag{13.24}$$

On using the regression theorem we have from (13.23)

$$\langle S^+(\tau_1) S^-(\tau_2)\rangle = \left\{ L_{\tau_1 - \tau_2}^{-1} \sum_m (v_{m+1})^{\frac{1}{2}} \sum_{l \geq 0} \left(\prod_1^l v_{m+k} v_{m+1+k}\right)^{\frac{1}{2}} \right.$$

$$\prod_0^l [z + \tfrac{1}{2}(\beta_+ v_{m+k} + \beta_- v_{m+k+1}) - i\omega']^{-1}(v_{m+l+1})^{-\frac{1}{2}} \right\}$$

$$\cdot \langle S^+|l+m\rangle \langle l+m|S^-\rangle_{\tau_2}$$

$$= \left\{ L_{\tau_1 - \tau_2}^{-1} \sum_m (v_{m+1})^{\frac{1}{2}} \sum_{l \geq 0} \left(\prod_1^l v_{m+k} v_{m+1+k}\right)^{\frac{1}{2}} \prod_0^l [z + \tfrac{1}{2}(\beta_+ v_{m+k} \right.$$

$$\left. + \beta_- v_{m+k+1}) - i\omega']^{-1}(v_{m+l+1})^{-\frac{1}{2}} \right\} (v_{m+l+1}) \varrho_{l+m+1, l+m+1}(\tau_2) \tag{13.25}$$

$$= \left\{ L_{\tau_1 - \tau_2}^{-1} \sum_m (v_{m+1})^{\frac{1}{2}} \sum_{l \geq 0} \left(\prod_1^l v_{m+k} v_{m+1+k}\right)^{\frac{1}{2}} \prod_0^l [z + \tfrac{1}{2}(\beta_+ v_{m+k} \right.$$

$$\left. + \beta_- v_{m+k+1}) - i\omega']^{-1}(v_{m+l+1})^{\frac{1}{2}} \right\} \left\{ L_{\tau_2}^{-1} \sum_{n \geq 0} \prod_{r=1}^n v_{m+l+1+r} \right.$$

$$\prod_0^n (z + v_{m+l+1+r})^{-1} \varrho_{m+l+1+n, m+l+1+n}(0) \Big\},$$

where we used (13.14). (13.25) is the exact result for the two-time correlation functions and it is quite involved. In particular, for the two-atom problem with each atom initially excited $[\varrho(0) = |1, 1\rangle \langle 1, 1|$ (Dicke state)] we find from (13.25) $[v_1 = v_0 = 2, v_{-1} = 0]$

$$\langle S^+(\tau_1) S^-(\tau_2)\rangle = e^{i\omega'(\tau_1 - \tau_2)}\{2e^{-2\tau_1}(1 - 2/\beta_-) + 4(\tau_2 + 1/\beta_+)$$

$$\cdot e^{-\beta_- \tau_1 - \tau_2(2 - \beta_-)}\}. \tag{13.26}$$

Such two-time correlation functions are useful in calculating the normally ordered correlation functions for the field operators, e.g. from (7.13) we find

$$E^{(+)}(r, t) \sim E_0^{(+)}(r, t) - k_0^2 \frac{\hat{r} \times (\hat{r} \times d)}{r} S^-(t - r/c),$$

and therefore

$$\langle E^{(-)}(r_1, t_1) E^{(+)}(r_2, t_2)\rangle = k_0^4 \frac{(\hat{r}_1 \times (\hat{r}_1 \times d)) (\hat{r}_2 \times (\hat{r}_2 \times d))}{r_1 r_2}$$

$$\langle S^+(t_1 - r_1/c) S^-(t_2 - r_2/c)\rangle. \tag{13.27}$$

From (7.27) and (13.25) we obtain the mean number of photons in any mode ks and its steady-state value as

$$\hat{N}_{ks}(z) = z^{-1}|g_{ks}|^2 \sum_m (v_{m+1})^{\frac{1}{2}} \sum_{l \geq 0} \left(\prod_1^l v_{m+k} v_{m+1+k}\right)^{\frac{1}{2}}$$

$$\cdot \prod_0^l [z + i\omega_{ks} + \tfrac{1}{2}(\beta_+ v_{m+k} + \beta_- v_{m+k+1}) - i\omega']^{-1}(v_{m+l+1})^{\frac{1}{2}} \tag{13.28}$$

$$\cdot \sum_{n \geq 0} \prod_{r=1}^n v_{m+l+1+r} \prod_0^n (z + v_{m+l+1+r})^{-1} \varrho_{m+l+1+n.\,m+l+1+n}(0) + \text{H.C.},$$

$$N_{ks}(\infty) = |g_{ks}|^2 \sum_m (v_{m+1})^{\frac{1}{2}} \sum_{l \geq 0} \left(\prod_1^l v_{m+k} v_{m+1+k}\right)^{\frac{1}{2}}$$

$$\prod_0^l [i\omega_{ks} - i\omega' + \tfrac{1}{2}(\beta_+ v_{m+k} + \beta_- v_{m+k+1})]^{-1}(v_{m+l+1})^{-\frac{1}{2}} \tag{13.29}$$

$$\cdot \sum \varrho_{m+l+1+n.\,m+l+1+n}(0) + \text{H.C.}$$

Equation (13.29) gives us the line shape of the radiation emitted from a small sample of identical two-level atoms. Again for a system of two two-level atoms with each atom initially in its excited state, we obtain for the line shape

$$\langle a_{ks}^+ a_{ks}\rangle \xrightarrow{t \to \infty} \frac{2|g_{ks}|^2 (X^2 + 40\gamma^2)}{(X^2 + 16\gamma^2)(X^2 + 4\gamma^2)}, \qquad X = \omega - \omega_{ks}, \tag{13.30}$$

and we have ignored the frequency-shift terms α_+. The result (13.30) is in agreement with the result obtained by using the Goldberger-Watson method.

We now consider equations of motion for the collective operators as well as the individual atomic operators, and we present the Langevin equations. In what follows we also ignore the frequency-shift terms so that we are working effectively with the master equation

$$\partial\varrho/\partial t = -i\omega[S^z, \varrho] - \gamma(S^+ S^- \varrho - 2S^- \varrho S^+ + \varrho S^+ S^-)$$
$$= -i\omega \sum_i [S_i^z, \varrho] - \gamma \sum_{ij} (S_i^+ S_j^- \varrho - 2S_j^- \varrho S_i^+ + \varrho S_i^+ S_j^-). \tag{13.31}$$

It is easy to show from (13.31) that $\langle S_i \cdot S_j \rangle$ is a constant of motion due to the permutation symmetry in the problem; if the atoms are initially excited to a permutationally symmetric state, then

$$\langle S_i \cdot S_j \rangle = \tfrac{1}{4} \quad (i \neq j). \tag{13.32}$$

From (13.31) we find that

$$(\partial/\partial t) \langle S_i^z \rangle = -2\gamma \langle \tfrac{1}{2} + S_i^z \rangle - \gamma \sum_{j \neq i} (\langle S_i^+ S_j^- \rangle + \langle S_j^+ S_i^- \rangle),$$

which on using (13.32) reduces to

$$(\partial/\partial t) \langle S_i^z \rangle = -2\gamma \langle \tfrac{1}{2} + S_i^z \rangle - \gamma \left\{ \tfrac{1}{2}(N-1) - 2 \sum_{j \neq i} \langle S_i^z S_j^z \rangle \right\}.$$

Since we are dealing with a small sample for which permutation symmetry exists, it follows that the mean values like $\langle S_i^z \rangle$ and $\langle S_i^z S_j^z \rangle$ would be same for any pair of atoms. Hence from the above we find that the total energy of the atomic system in units of ω obeys the equation

$$\partial W/\partial t + 2\gamma(W + \tfrac{1}{2}N) + 2\gamma(N-1)\{W + \tfrac{1}{2}N - N\langle S_i^+ S_i^- S_j^+ S_j^- \rangle\} = 0$$
$$(i \neq j). \tag{13.33}$$

On the other hand, we find from (13.31) that $\langle S_i^+ S_j^+ S_i^- S_j^- \rangle$ obeys [cf. (8.23)]

$$(\partial/\partial t) \langle S_i^+ S_j^+ S_i^- S_j^- \rangle + 4\gamma \langle S_i^+ S_j^+ S_i^- S_j^- \rangle$$
$$+ \gamma \sum_{l \neq i \neq j} \{\langle S_i^+ S_j^+ S_i^- S_j^- \rangle + \langle S_i^+ S_j^+ S_i^- S_l^- \rangle + \text{c.c.}\} = 0 \quad (i \neq j). \tag{13.34}$$

For the collective operators one has

$$(\partial/\partial t) \langle S^z \rangle = -2\gamma \langle S^+ S^- \rangle,$$
$$(\partial/\partial t) \langle S^+ \rangle = 2\gamma \langle S^+ S^z \rangle + i\omega \langle S^+ \rangle,$$
$$(\partial/\partial t) \langle S^- \rangle = 2\gamma \langle S^z S^- \rangle - i\omega \langle S^- \rangle, \tag{13.35}$$

$$(\partial/\partial t) \langle S^+ S^- \rangle = 4\gamma \langle S^+ S^z S^- \rangle, \tag{13.36}$$

etc. We note here that if we make *a priori* a semiclassical approximation, i.e. factorize the mean values $\langle S^+ S^z \rangle \approx \langle S^+ \rangle \langle S^z \rangle$ etc., then we find that (13.35) reduce to the neoclassical equations of motion (Chapter 16). We will also see in Chapter 16 that such a factorization cannot be done, even *a priori* for a single two-level atom because of the intrinsic property of spin-$\frac{1}{2}$ angular momentum operators. The nonlinear Langevin equations in terms of the collective operators are

$$\dot{S}^+ = i\omega S^+ + 2\gamma S^+ S^z + F^+ , \qquad \dot{S}^- = -i\omega S^- + 2\gamma S^z S^- + F^- , \qquad (13.37)$$

where the random force $F^\pm$ has the property

$$\langle F^+(t) F^-(t') \rangle = \langle F^+(t) F^+(t') \rangle = \langle F^-(t) F^-(t') \rangle = 0 ,$$
$$\langle F^-(t) F^+(t') \rangle = 2 \langle D^{-+} \rangle \delta(t - t') , \qquad (13.38a)$$

$\langle D^{-+} \rangle$ being obtained from the Einstein relation

$$2 \langle D^{-+} \rangle = 8\gamma \langle S^z(t) S^z(t) \rangle . \qquad (13.38b)$$

In the next section we will discuss the c-number Langevin equations.

In order to calculate the radiation rate $I(t)$ ($\propto \partial W / \partial t$), we see that we should solve (13.33), which contains a two-particle mean value. The equations of motion for the two-particle mean values in turn contain three-particle mean values and so on, hence one obtains a whole hierarchy of equations. We have already given an expression for the radiation rate [(13.17) with $q = 0$, $p = 1$] in terms of the solution (13.14) of the master equation. The solution is obviously quite an involved one; when the values of the number of atoms are large, we can either obtain the exact solution on a computer, or we can proceed by making suitable approximations (to an accuracy of order $1/N$). Some of the approximate methods for calculating the radiation rate are discussed at length in [29, 31]. The results of numerical computations are presented in [78].

The approximations used in calculating the radiation rate center about the two-particle mean value $\langle S_i^+ S_j^+ S_i^- S_j^- \rangle$ which appears in (13.33). The obvious thing to do is to express the two-particle mean value in terms of the one-particle mean values. We can make either of the approximations

(I) $\langle S_i^+ S_j^+ S_i^- S_j^- \rangle \approx \langle S_i^+ S_i^- \rangle \langle S_j^+ S_j^- \rangle , \qquad (i \neq j) ,$

(II) $\langle S_i^+ S_j^+ S_i^- S_j^- \rangle \approx \langle S_i^+ S_i^- \rangle \langle S_j^+ S_j^- \rangle + \langle S_i^+ S_j^- \rangle \langle S_j^+ S_i^- \rangle , \qquad (i \neq j)$

depending on the initial condition, because the chosen approximation has to be consistent with it. We have elsewhere referred to I and II as the Hartree and Hartree-Fock approximations (time-dependent), respectively [66]. The Hartree-Fock approximation is in a sense a nonlinear type of decoupling scheme. We now discuss some initial excitations.

(I) Initial Θ Excitation $(\theta < \pi)$ Eq. (8.4)

In this case only the Hartree approximation is consistent with the initial condition and then (13.33) leads to

$$\partial W/\partial t = 2\gamma(W + \tfrac{1}{2}N)\{(1 - 1/N)(W - \tfrac{1}{2}N) - 1\}, \tag{13.39}$$

which is easily integrated, whereupon the radiation rate is found to be

$$I(t) = \tfrac{1}{2}\omega\gamma N^3(N - 1)^{-1} \operatorname{sech}^2\{N\gamma(t - \tau)\}$$
$$\tau = (2N\gamma)^{-1}\ln(N - 1)[1 + N\cot^2(\theta_0/2)]^{-1}, \tag{13.40}$$

which is the well-known behavior of the radiation rate. A similar result has been obtained by Rehler and Eberly using very different kind of approach, with the difference that we are dealing with small samples whereas Rehler and Eberly [49] discuss the case of extended systems. We have also discussed at length in [31] the accuracy of the Hartree approximation, whence we find that the correction terms to it are of the order $1/N$. It should also be noted that to use the Hartree approximation on the two-particle mean value $\langle S_i^+ S_j^+ S_i^- S_j^- \rangle$ does not mean that the density operator of the N-particle system can be written as a product of one-particle density matrices. Furthermore, for the Θ excitation the dipole moment possesses a macroscopic value.

(II) Initial Excitation to Dicke State $|\tfrac{1}{2}N, \tfrac{1}{2}N\rangle$

Next we consider the case where the system was initially prepared in a totally inverted state. In this case both the Hartree and Hartree-Fock approximations are consistent with the initial condition. We have seen above [Eq. (13.15b)] that the dipole moment of the system remains zero and hence any superradiant emission will be due to the presence of correlations, so that the Hartree approximation would be a poor choice. Support for this view is also found if we recall that for the harmonic oscillator model correlations were induced among different oscillators when the system was initially excited to a Fock state. We therefore adopt the Hartree-Fock scheme which in conjunction with (13.32) leads to

$$\langle S_i^+ S_i^- S_j^+ S_j^- \rangle \approx \langle S_i^+ S_i^- \rangle + \tfrac{1}{2} - \{\tfrac{1}{4} + \langle S_i^+ S_i^- \rangle - \langle S_i^+ S_i^- \rangle^2\}^{\tfrac{1}{2}}. \tag{13.41}$$

On combining (13.41) and (13.33) we obtain

$$\partial W/\partial t + 2\gamma(W + \tfrac{1}{2}N) - \gamma N(N - 1) + 2\gamma N(N - 1)\{\tfrac{1}{2} - W^2/N^2\}^{\tfrac{1}{2}} = 0, \tag{13.42}$$

the solution of which is

$$-b(b^2+c^2)^{-1}\ln(a+b\cos x+c\sin x)+c(b^2+c^2)^{-1}x-ac(b^2+c^2)^{-1}$$

$$\cdot(b^2+c^2-a^2)^{-\frac{1}{2}}$$

$$\cdot\ln\left|\frac{(a-b)\tan(x/2)+c-(b^2+c^2-a^2)^{\frac{1}{2}}}{(a-b)\tan(x/2)+c+(b^2+c^2-a^2)^{\frac{1}{2}}}\right|=2\gamma(t-t_0), \tag{13.43a}$$

$$a=-N+2/\sqrt{2}, b=1, c=(N-1), W=(N/\sqrt{2})\cos x, \tag{13.43b}$$

with t_0 determined from the initial condition and for large N (for which all the approximations are expected to hold)

$$\left|\frac{\tan(x/2)-\alpha}{\tan(x/2)-\beta}\right|=\exp\{2N\gamma(t-t_0)-x\},$$

$$\alpha=\sqrt{2}-1-2\sqrt{2}N^{-1}(\sqrt{2}-1), \quad \beta=(\sqrt{2}+1). \tag{13.44}$$

The radiation rate from (13.42) is

$$I(t)=2\gamma\omega[W+N-\tfrac{1}{2}N^2+(N-1)(\tfrac{1}{2}N^2-W^2)^{\frac{1}{2}}]. \tag{13.45}$$

Near the point of superradiant emission $W\approx0$,

$$I=2\gamma\omega(N^2/4)(0.828), \tag{13.46}$$

and also

$$\langle S_i^z S_j^z\rangle-\langle S_i^z\rangle\langle S_j^z\rangle\approx0.045. \tag{13.47}$$

These results have been obtained by the use of very different arguments in [58] [following Eq. (155)] and [79]. The radiation rates as predicted by (13.40) and (13.45) agree very well with the numerical solution of the master equation [33].

Finally, it should be admitted that we have treated the dipole-dipole coupling that appears in (13.1) rather unsystematically; this has been the subject of some recent investigations [85]. Dicke in his original work assumed that the atoms are so far apart that the dipole-dipole coupling is negligible. We have seen that the radiation characteristics of a small sample are determined by the collective operators, and hence it seems reasonable that the radiation properties will be on the average independent of the exact location of the atoms in a small sample. For this reason we can introduce a density operator $\bar{\varrho}$ averaged over the position of the atoms and then project out from the master equation (13.1) an equation of motion for the averaged density operator. It is easily seen from (13.1) that in the lowest order in the V_{ij} the equation of motion for $\bar{\varrho}$ is

$$\partial\bar{\varrho}/\partial t = -i\omega_0[S^z, \bar{\varrho}] - \gamma(S^+ S^- \bar{\varrho} - 2S^- \bar{\varrho} S^+ + \bar{\varrho} S^+ S^-)$$
$$-iv[S^+ S^- - S^z, \bar{\varrho}] = -i\omega_0[S^z, \bar{\varrho}] - iv[S^+ S^-, \bar{\varrho}] \qquad (13.48)$$
$$-\gamma(S^+ S^- \bar{\varrho} - 2S^- \bar{\varrho} S^+ + \bar{\varrho} S^+ S^-),$$

where in passing from the first to the second line we absorbed v in the definition of ω_0 and v denotes the ensemble average of V_{ij} over the distribution of atoms in the small sample. The structure of the master equation (13.48) is similar to that of (13.5). It should be noted that (13.5) can be rewritten in the form

$$\partial\varrho/\partial t = -i(\omega + \Omega_{ii})[S^z, \varrho] - i\Omega_{ii}^-[S^+ S^- - S^z, \varrho]$$
$$-\gamma(S^+ S^- \varrho - 2S^- \varrho S^+ + \varrho S^+ S^-), \qquad (13.49)$$

where

$$\Omega_{ii} = -(\alpha_+ - \alpha_-), \qquad \Omega_{ii}^- = -(\alpha_+ + \alpha_-), \qquad (13.50)$$

which coincide with the frequency shifts (6.47) and (6.66), respectively. In Dicke's model the parameter Ω_{ii}^- essentially plays the role of the average of V_{ij}. It is evident from (13.48) that the diagonal matrix elements of $\bar{\varrho}$ satisfy the same Eq. as (13.9). Since the radiation rate is determined only from the diagonal elements [cf. (13.17)], it is clear that the radiation rate is unaffected[21]. From (13.48) we find that the collective operators $S^\pm$, S^z satisfy the equations

$$(\partial/\partial t)\langle S^+\rangle = i\omega_0\langle S^+\rangle + 2\gamma\langle S^+ S^z\rangle - 2iv\langle S^+ S^z\rangle,$$
$$(\partial/\partial t)\langle S^z\rangle = -2\gamma\langle S^+ S^-\rangle, \qquad (13.51)$$

and if one makes a semiclassical approximation, then

$$(\partial/\partial t)\langle S^\pm\rangle = \pm i\omega_0\langle S^\pm\rangle + 2\gamma\langle S^\pm\rangle\langle S^z\rangle \mp 2iv\langle S^\pm\rangle\langle S^z\rangle,$$
$$(\partial/\partial t)\langle S^z\rangle = -2\gamma\langle S^+\rangle\langle S^-\rangle. \qquad (13.52)$$

Equations (13.52) coincide with the equations obtained by Stroud et al. [86] using the neoclassical theory.

---

[21] Although the radiation rate on the average is independent of V_{ij}, the single atom expectation values would depend on V_{ij}.

14. Emission from a Small Sample of Two-Level Atoms: Approximate Solution of the Master Equation and the Langevin Equation

In the previous section we presented the exact solution of the master equation as well as some approximate results for the radiation rate. As already remarked, the exact solution of the master equation was quite involved and for practical purposes it seems necessary to resort to approximate methods. In the present section we will discuss the c-number Langevin equations, which can also be solved exactly, and from the exact solution we will obtain a number of approximate solutions.

The Langevin equations describing the spontaneous emission from a collection of two-level atoms confined to a region smaller than a wavelength are from (8.19) (ignoring the cooperative frequency shifts)

$$\dot{z}_i = -(i\omega_0 + \gamma) z_i + \gamma \sum_{j\neq i} z_j(2|z_i|^2 - 1), \quad i = 1, 2, \dots, N. \tag{14.1}$$

The fact that the random force is absent from (14.1) does not mean that there are no atomic correlations. z_i are fluctuating quantities because of the initial condition. If we define

$$D = \sum_i z_i, \quad J = DD^* = \psi^{-2}, \quad D = J^{\frac{1}{2}} e^{i\varphi}, \quad A = \sum_i |z_i|^2, \tag{14.2}$$

then from (14.1) we find

$$\dot{D} = -i\omega_0 D - N\gamma D + 2\gamma DA, \tag{14.3}$$

$$\dot{A} = -2\gamma DD^*, \tag{14.4}$$

and from these we easily find

$$\varphi(t) = -i\omega_0 t + \varphi_0, \tag{14.5}$$

$$(\partial^2/\partial\tau^2) \ln J + 2J = 0, \quad \tau = 2\gamma t. \tag{14.6}$$

The solution of (14.6) is

$$J = \tfrac{1}{4} a \operatorname{Sech}^2 \tfrac{1}{2}\sqrt{a}\,(\tau - \tau_0), \tag{14.7}$$

where a and τ_0 are the parameters depending on the initial condition

$$J_0^{-\frac{1}{2}} = 2a^{-\frac{1}{2}} \cosh \tfrac{1}{2}\sqrt{a}\,\tau_0, \quad a = 4J_0 + (2A_0 - N)^2. \tag{14.8}$$

The time dependence of A is given by

$$A = \tfrac{1}{2}N - \tfrac{1}{2}\sqrt{a} \tanh \tfrac{1}{2}\sqrt{a}\,(\tau - \tau_0). \tag{14.9}$$

The quantities J_0, a, τ_0 and A_0, being the values which the Langevin variables take at $t = 0$, are not fixed numbers but fluctuating variables. Because we used the normal ordering in obtaining (14.1), the mean value of (14.7) gives us

$$\langle J \rangle = \sum_{ij} \langle z_i^* z_j \rangle = \sum_{ij} \langle S_i^+ S_j^- \rangle = \langle S^+ S^- \rangle$$
$$= \langle \tfrac{1}{4} a \operatorname{Sech}^2 \tfrac{1}{2} \sqrt{a} \, (\tau - \tau_0) \rangle , \tag{14.10}$$

where the average on the right-hand side refers to the initial distribution. The radiation rate $I(t)$ will be

$$I(\tau) = 2\gamma\omega\langle S^+ S^- \rangle = 2\gamma\omega\langle \tfrac{1}{4} a \operatorname{Sech}^2 \tfrac{1}{2} \sqrt{a} \, (\tau - \tau_0) \rangle . \tag{14.11}$$

It is interesting to note that the radiation rate is given by the average of a "Sech" solution. In the solution of the Langevin equations we see that two random variables occur, viz. a and τ_0, and thus the averaging in (14.11) can be replaced by the averaging with reference to the distribution of a and J_0. We first examine the behavior of the random variable a. It is clear from (14.8) and our normal ordering rule that

$$\langle a \rangle = 4 \sum_{ij} \langle S_i^+ S_j^- \rangle + 4 \sum_{ij} \langle S_i^+ S_j^+ S_i^- S_j^- \rangle + N^2 - 4N \sum_i \langle S_i^+ S_i^- \rangle$$

$$= 4\langle S^+ S^- \rangle + N^2 - 4N(\langle S^z \rangle + \tfrac{1}{2}N) + 4\langle (S^z + \tfrac{1}{2}N)^2 \rangle - 4\langle S^z + \tfrac{1}{2}N \rangle ,$$

or on simplification

$$\langle a \rangle = N^2 , \tag{14.12}$$

where we made use of the relation $S^+ S^- = \tfrac{1}{2}N(\tfrac{1}{2}N + 1) - S^z S^z + S^z$. Furthermore, if the normal ordering is ignored, it is clear that $a \approx 4S^+ S^- + 4S^z S^z = 4\{\tfrac{1}{2}N(\tfrac{1}{2}N + 1) + S^z\}$ and thus the fluctuations in a will be due to the fluctuations in S^z (as well as to the normal ordering). Therefore we can write

$$a = N^2 + O(N) , \tag{14.13}$$

and up to terms of $O(1/N)$, a can be regarded as a deterministic variable. Then we have

$$J \approx \tfrac{1}{4}N^2 \operatorname{Sech}^2 \tfrac{1}{2}N(\tau - \tau_0) , \qquad A \approx \tfrac{1}{2}N\{1 - \tanh \tfrac{1}{2}N(\tau - \tau_0)\} ,$$
$$\cosh \tfrac{1}{2}N\tau_0 \approx \tfrac{1}{2}N J_0^{-\frac{1}{2}} . \tag{14.14}$$

The fluctuating variable is J_0 or τ_0 and hence there is a kind of time zitter [87, 88]. The distribution of J_0 depends on the initial condition. The moments of J_0 are given by

$$\langle J_0^n \rangle = \langle (S^+)^n (S^-)^n \rangle$$

and the distribution of J_0 is given by

$$P(J_0) = \sum_0^\infty \frac{(-1)^n}{n!} \langle (S^+)^n (S^-)^n \rangle \, \delta^{(n)}(J_0). \tag{14.15}$$

For the case when the system was initially excited to a Dicke state $|\frac{1}{2}N, m\rangle$, $P(J_0)$ reduces to

$$P(J_0) = \sum_0^{m+\frac{1}{2}N} \frac{(\frac{1}{2}N+m)! \, (\frac{1}{2}N-m+n)!}{(\frac{1}{2}N+m-n)! \, (\frac{1}{2}N-m)!} \frac{(-1)^n}{n!} \, \delta^{(n)}(J_0). \tag{14.16}$$

The behavior of this distribution for different regions of m values has been studied by Haake and Glauber [80]; using the approximation $(x+\alpha)!/x! \approx x^\alpha$ for large x, they find that

$$P(J_0) = \delta(J_0 - \tfrac{1}{4}N^2 + m^2) \quad \text{for} \quad m \ll \tfrac{1}{2}N, \tag{14.17}$$

$$= \frac{1}{l!} \left(\frac{J_0}{N-l} \right)^l (N-l)^{-1} e^{-J_0/(N-l)} \quad \text{for} \quad \tfrac{1}{2}N - m \equiv l \ll \tfrac{1}{2}N. \tag{14.18}$$

We first consider the region $m \ll \tfrac{1}{2}N$. In this case the probability distribution of J_0 is a delta function which implies that there are *no fluctuations*. The radiation rate will be on combining (14.17), (14.13) and (14.11)

$$I(\tau) \approx 2\gamma\omega(\tfrac{1}{4}N^2) \, \text{Sech}^2 \, \tfrac{1}{2}N(\tau - \tau_0), \quad \tau_0 = 2N^{-1} \cosh^{-1} \tfrac{1}{2}N(\tfrac{1}{4}N^2 - m^2)^{-\frac{1}{2}}. \tag{14.19}$$

The energy of the atomic system is

$$\langle S^z(\tau) \rangle = \sum_i \langle S_i^+ S_i^- \rangle - \tfrac{1}{2}N = \langle A \rangle - \tfrac{1}{2}N = -\tfrac{1}{2}N \tanh \tfrac{1}{2}N(\tau - \tau_0), \tag{14.20}$$

and indeed, normally ordered correlations are given by

$$\langle (S^+)^n (S^z)^k (S^-)^n \rangle \approx [\tfrac{1}{4}N^2 \, \text{Sech}^2 \, \tfrac{1}{2}N(\tau - \tau_0)]^n$$
$$\cdot [-\tfrac{1}{2}N \tanh \tfrac{1}{2}N(\tau - \tau_0)]^k \quad n < m \ll \tfrac{1}{2}N. \tag{14.21}$$

It should be noted that the above results are also obtained directly from the equations of motion (13.35) and (13.36) if one makes a semiclassical approximation $\langle (S^+)^n (S^z)^k (S^-)^n \rangle \approx \langle S^+ S^- \rangle^n \langle S^z \rangle^k$. The two-time correlation function will be

$$\langle S^+(\tau_1) S^-(\tau_2) \rangle \approx e^{i\omega_0(\tau_1 - \tau_2)} (\tfrac{1}{4}N^2) \, \text{Sech} \, \tfrac{1}{2}N(\tau_1 - \tau_0) \, \text{Sech} \, \tfrac{1}{2}N(\tau_2 - \tau_0), \tag{14.22}$$

which in turn determines the normally ordered correlation function for the field operator.

For the case $\frac{1}{2}N - m \equiv l \ll \frac{1}{2}N$, with l being a small number, one can write approximately $e^{N\tau_0/2} \approx N/\sqrt{J_0}$ and then on combining (14.18), (14.13) and (14.11) we obtain for the radiation rate

$$I(t) = 2\gamma\omega \int_0^\infty dJ_0 (1/l!) \left(\frac{J_0}{N}\right)^l N^{-1} e^{-J_0/N} J_0 e^{2N\gamma t}$$
$$\cdot \{1 + (J_0/N^2) e^{2N\gamma t}\}^{-2} . \tag{14.23}$$

Any normally ordered correlation of the form $\langle (S^+)^n (S^z)^k (S^-)^n \rangle$ reduces to

$$\langle (S^+)^n (S^z)^k (S^-)^n \rangle \approx \int_0^\infty dJ_0 (1/l!) (J_0/N)^l N^{-1}$$
$$\cdot e^{-J_0/N} \{J_0 e^{2N\gamma t}/(1 + (J_0/N^2) e^{2N\gamma t})^2\}^n \left\{\frac{1}{2}N\left(\frac{1 - (J_0/N^2) e^{2N\gamma t}}{1 + (J_0/N^2) e^{2N\gamma t}}\right)\right\}^k . \tag{14.24}$$

The two-time correlation function is given by

$$\langle S^+(t_1) S^-(t_2) \rangle \approx \int_0^\infty dJ_0 (1/l!) (J_0/N)^l N^{-1} e^{-J_0/N} J_0 e^{N\gamma(t_1 + t_2)}$$
$$\cdot e^{i\omega_0(t_1 - t_2)} \{(1 + (J_0/N^2) e^{2N\gamma t_1})(1 + (J_0/N^2) e^{2N\gamma t_2})\}^{-1} . \tag{14.25}$$

A result of the form (14.24) for the case $k = 0$, $l = 0$ was first obtained by Degiorgio [87], who invoked more or less on phenomenological grounds (cf. [88]) the distribution (14.18) for the initial radiation rate. These results were subsequently derived by Haake and Glauber in Ref. [80], using a different method on which we shall comment later. The above results can be expressed in terms of the exponential integrals [89]

$$E_1(a) \equiv \int_a^\infty (dz/z) e^{-z} . \tag{14.26}$$

We have, for example, for the case $l = 0$, i.e. for the completely inverted state $|\frac{1}{2}N, \frac{1}{2}N\rangle$

$$I(t) = N^2(1 + a^{-1} \partial/\partial a) [a^{-1} e^{1/a} E_1(1/a)], \quad a = N^{-1} e^{2N\gamma t}, \tag{14.27}$$

$$\langle S^+(t_1) S^-(t_2) \rangle = \frac{N e^{i\omega(t_1 - t_2) + N\gamma(t_1 + t_2)}}{(a_2 - a_1)} \{a_1^{-1} e^{a_1^{-1}} E_1(a_1^{-1})$$
$$- a_2^{-1} e^{a_2^{-1}} E_1(a_2^{-1})\}, \quad a_i = N^{-1} e^{2N\gamma t_i} . \tag{14.28}$$

We will conclude this section by briefly commenting on the method used by Haake and Glauber. They introduced the characteristic function

$$C(\xi, \xi^*, \zeta, t) \equiv \langle e^{i\xi S^+} e^{i\zeta S^z} e^{i\xi^* S^-} \rangle, \tag{14.29}$$

and the corresponding distribution function

$$P(s, s^*, m, t) = (1/2\pi^3) \int \int d^2\xi \, d\zeta \, e^{-i\xi s^* - i\xi^* s - i\zeta m} \, C(\xi, \xi^*, \zeta, t). \tag{14.30}$$

Using the master Eq. (13.1) (with the effect V_{ij} ignored), a straightforward calculation shows that

$$\partial P/\partial \tau = \left\{ -\left(\frac{\partial}{\partial s^*} s^* + \frac{\partial}{\partial s} s \right) m + (1 - e^{-\partial/\partial m}) ss^* \right.$$
$$\left. + \frac{1}{2} \left(\frac{\partial^2}{\partial s^{*2}} s^{*2} + \frac{\partial^2}{\partial s^2} s^2 \right) \right\} P, \tag{14.31}$$

which contains derivatives with respect to m to all orders. Equation (14.31) is equivalent to the Langevin equations

$$\dot{s} = ms + F_s, \quad \dot{s}^* = ms^* + F_s^*, \quad \dot{m} = -ss^* + F_m, \tag{14.32}$$

where the random forces F_s, F_m are delta-correlated but are not Gaussian. The correlation functions of these random forces are extremely involved and are functionals of the variables s, s^* and m. Haake and Glauber ignored the fluctuating forces in (14.32) and solved the deterministic equations. The randomness is introduced through the inital condition and they found the results (14.16) to (14.24) for large values of N. They later argued that the effect of the fluctuating forces is to introduce correction terms of order $1/N$, and hence is negligible. In our approach to results (14.16) to (14.28), we solved the Van der Pol type of equations (possible due to the intrinsic nature of spin-$\frac{1}{2}$ systems) which involve no random forces for spontaneous emission because we adopted the normal ordering rule. Our solution involved two fluctuating parameters a and τ_0. The effect of using the fluctuating variable τ_0 is identical to that of retaining the deterministic part in (14.32). It is suspected that the role of the variable a, which was found to a good approximation (to order $1/N$) to be deterministic (nonstochastic), must be similar to the role of the random forces in the Langevin equations (14.32).

15. Master-Equation Treatment of Spontaneous Emission from a Multilevel Atom

Let us consider the spontaneous emission from a multilevel atom using the master-equation approach. We discussed spontaneous emission from a three-level atom in Chapter 3 and Chapter 5 using the Weisskopf-Wigner and Goldberger-Watson methods. We first treat the case of an atom having a nonequidistant, nondegenerate spectrum. We write the interaction Hamiltonian in the form

$$H = \sum_j E_j A_{jj} + \sum_{ks} \omega_{ks} a_{ks}^+ a_{ks} + \sum_{kl} v_{kl} A_{kl}, \tag{15.1}$$

where

$$v_{il} = i \sum_{ks} (2\pi ck/L^3)^{\frac{1}{2}} (d_{il} \cdot \varepsilon_{ks}) a_{ks} - i \sum_{ks} (2\pi ck/L^3)^{\frac{1}{2}} (d_{il} \cdot \varepsilon_{ks}^*) a_{ks}^+, \tag{15.2}$$

$$= 0 \quad \text{if} \quad i = l.$$

The master-equation for the reduced-density operator corresponding to the atomic system can be obtained by using the formalism of Chapter 6. The derivation is reviewed in several places [27, 52, 90] so here we merely quote the result obtained by making Born, Markov and rotating-wave approximations,

$$\partial \varrho / \partial t = \sum_{klmn} \{ (A_{mn} \varrho A_{kl} - A_{kl} A_{mn} \varrho) \gamma_{klmn}^+ \tag{15.3}$$

$$+ (A_{kl} \varrho A_{mn} - \varrho A_{mn} A_{kl}) \gamma_{mnkl}^- \},$$

where the summation is over those values of k, l, m, n which satisfy

$$k = n, m = l, \quad k = l, m = n, \quad k = l = m = n. \tag{15.4}$$

On transforming to the Schrödinger picture we obtain after a number of simplifications

$$\partial \varrho_{ij} / \partial t = - i \omega_{ij} \varrho_{ij} - \Gamma_{ji} \varrho_{ij} + \delta_{ij} \sum_{k \neq i} 2 \gamma_{ik} \varrho_{kk}, \tag{15.5}$$

where

$$\omega_{ij} = E_i - E_j, \quad \operatorname{Re} \Gamma_{ji} = \tfrac{1}{2} (\Gamma_i + \Gamma_j), \quad \Gamma_i = \sum_{k \neq i} 2 \gamma_{ki},$$

$$2 \gamma_{lk} = \int_{-\infty}^{+\infty} \langle v_{kl}(t) v_{lk}(0) \rangle e^{-i\omega_{lk}t} dt. \tag{15.6}$$

Here $2\gamma_{lk}$ is the transition probability per unit time that the system makes a transition from the state $|k\rangle$ to $|l\rangle$ as a result of spontaneous emission. Using (15.2) we find that

$$\gamma_{ij} = 0 \quad \text{if} \quad E_i > E_j,$$

$$= \tfrac{2}{3} |d_{ij}|^2 / c^{-3} \omega_{ji}^3 \quad \text{if} \quad E_j > E_i. \tag{15.7}$$

Thus $2\gamma_{ij}$, for $E_i < E_j$, is equal to the Einstein A coefficient, Γ_i is equal to the total transition probability that the system will make a transition from state i to all other states. The imaginary part of Γ_{ij} causes the level shifts and is given by

$$\text{Im}\,\Gamma_{ij} \equiv -\Delta_{ij} = (1/2i) \sum_l \int_0^\infty dt\,[\langle v_{jl}(t)\,v_{lj}(0)\rangle\,e^{-i\omega_{lj}t} \tag{15.8a}$$

$$+ \langle v_{il}(0)\,v_{li}(t)\rangle\,e^{-i\omega_{il}t} - \text{c.c.}] \tag{15.8b}$$

$$= \Delta_{ji},$$

which we will also write as

$$\Delta_{ij} = \Delta_{ii} - \Delta_{jj}, \tag{15.8c}$$

with

$$\Delta_{jj} = (1/2i) \sum_l \int_0^\infty dt\,[\langle v_{jl}(t)\,v_{lj}(o)\rangle\,e^{-i\omega_{lj}t} - \text{c.c.}] \tag{15.8d}$$

$$= -\sum_l \sum_{ks} |d_{jl} \cdot \varepsilon_{ks}|^2\,(2\pi ck/L^3)\,(\omega_{ks} + \omega_{lj})^{-1}, \tag{15.8e}$$

Δ_{jj} is the level shift of level j. The master equation (15.5) can now be written as

$$\partial \varrho_{ij}/\partial t = -i\omega'_{ij}\varrho_{ij} + \delta_{ij} \sum_{k \neq i} 2\gamma_{ik}\varrho_{kk} \tag{15.9}$$

$$- \left(\sum_{k \neq i} \gamma_{ki} + \sum_{k \neq j} \gamma_{kj}\right) \varrho_{ij},$$

where ω'_{ij} is the renormalized transition frequency

$$\omega'_{ij} = E'_i - E'_j = E_i + \Delta_{ii} - E_j - \Delta_{jj}. \tag{15.10}$$

The Lamb shifts, as given by (15.8e), are still to be renormalized, the $-d \cdot E$ interaction again picks up too many powers of k. We have already discussed this problem in detail in connection with a two-level atom. It is interesting to note that in the case of a two-level atom the change in the transition frequency as calculated by our approach did not contain any divergent terms. For a multi-level atom the situation is, of course, different. It should be further noted that the cubic divergence in (15.8e) is cancelled from the diagonal element of the transverse part of the contact term $2\pi \int |\boldsymbol{P}^\perp|^2\,d^3r$. The off-diagonal elements of the contact term correspond to counter-rotating terms which we have, of course, ignored in the derivation of (15.5).

The master equation (15.9) shows that the off-diagonal elements simply decay

$$\varrho_{ij}(t) = \exp\{-[i\omega_{ij} + \tfrac{1}{2}(\Gamma_i + \Gamma_j)]\,(t - t')\}\,\varrho_{ij}(t'), \quad (i \neq j), \tag{15.11}$$

where we have *suppressed the prime* from ω_{ij}, the decay constant being equal to half the sum of the decay constants of the ith and j^{th} levels. For the diagonal elements we must essentially solve the Pauli equation

$$\partial \varrho_{ii}/\partial t = - \sum_{k \neq i} 2\gamma_{ki} \varrho_{ii} + \sum_{k \neq i} 2\gamma_{ik} \varrho_{kk} \,. \tag{15.12}$$

One can rewrite (15.11) as

$$\langle A_{ji}(t) \rangle = \exp\{-[i\omega_{ij} + \tfrac{1}{2}(\Gamma_i + \Gamma_j)](t - t')\} \; \langle A_{ji}(t') \rangle, \quad (i \neq j), \tag{15.13}$$

and hence the two-time correlation function will be given by

$$\langle A_{ji}(t) A_{kl}(t') \rangle = \exp\{-[i\omega_{ij} + \tfrac{1}{2}(\Gamma_i + \Gamma_j)](t - t')\}$$
$$\cdot \langle A_{ji}(t') \rangle \, \delta_{ik}, \quad (i \neq j), \quad t > t' \,. \tag{15.14}$$

For a three-level atom in which the transitions $|1\rangle \rightarrow |2\rangle$ and $|2\rangle \rightarrow |3\rangle$ are allowed but $|1\rangle \rightarrow |3\rangle$ is forbidden $(E_1 > E_2 > E_3)$, (15.12) reduces to

$$\partial \varrho_{11}/\partial t = -2\gamma_{21} \varrho_{11},$$
$$\partial \varrho_{22}/\partial t = -2\gamma_{32} \varrho_{22} + 2\gamma_{21} \varrho_{11}, \tag{15.15}$$
$$\partial \varrho_{33}/\partial t = 2\gamma_{32} \varrho_{22},$$

which have the solutions

$$\varrho_{11}(t) = e^{-2\gamma_{21}t} \varrho_{11}(0),$$
$$\varrho_{22}(t) = e^{-2\gamma_{32}t} \varrho_{22}(0) + \varrho_{11}(0)\gamma_{21}(\gamma_{21} - \gamma_{32})^{-1} [e^{-2\gamma_{32}t} - e^{-2\gamma_{21}t}],$$
$$\varrho_{33}(t) = 1 - \varrho_{11}(t) - \varrho_{22}(t). \tag{15.16}$$

Let us now discuss the properties of the radiation field. By carrying out an analysis similar to that in Chapter 7, it can be shown that the positive-frequency part of the electric-field operator in the radiation zone is given by

$$E^{(+)}(r, t) \cong E_0^{(+)}(r, t) - (\omega_{12}^2/c^2 r)\, \hat{r} \times (\hat{r} \times d_{12})\, A_{21}(t - r/c) \tag{15.17}$$
$$- (\omega_{23}^2/c^2 r)\, \hat{r} \times (\hat{r} \times d_{23})\, A_{32}(t - r/c),$$

and hence the normally ordered correlation function is given by

$$\langle E^{(-)}(r_1, t_1) E^{(+)}(r_2, t_2) \rangle = (r_1 r_2)^{-1} (\hat{r}_1 \times (\hat{r}_1 \times d_{12}^*)) (\omega_{12}/c)^4$$
$$\cdot \langle A_{12}(t_1 - r_1/c) A_{21}(t_2 - r_2/c) \rangle (\hat{r}_2 \times (\hat{r}_2 \times d_{12})) \tag{15.18}$$

$+$ terms with $1 \rightarrow 2, 2 \rightarrow 3$,

which on using (15.14) and (15.16) reduces to

$$
\begin{aligned}
\langle E^{(-)}(r_1, t_1)\, E^{(+)}(r_2, t_2)\rangle &= (r_1 r_2)^{-1}\,(\hat{r}_1 \times (\hat{r}_1 \times d_{12}^*))\,(\hat{r}_2 \times (\hat{r}_2 \times d_{12}))\,(\omega_{12}/c)^4 \\
&\quad \cdot \exp\{[i\omega_{12} - (\gamma_{21} + \gamma_{32})]\,(t_1 - t_2 - r_1/c + r_2/c) - 2\gamma_{21}(t_2 - r_2/c)\} \\
&\quad + (r_1 r_2)^{-1}\,(\hat{r}_1 \times (\hat{r}_1 \times d_{23}^*))\,(\hat{r}_2 \times (\hat{r}_2 \times d_{23}))\,(\omega_{23}/c)^4\,[\gamma_{21}(\gamma_{21} - \gamma_{31})^{-1}] \\
&\quad \cdot \exp\{(i\omega_{23} - \gamma_{32})(t_1 - t_2 - r_1/c + r_2/c)\}\,\{e^{-2\gamma_{32}(t_2 - r_2/c)} - e^{-2\gamma_{21}(t_2 - r_2/c)}\}.
\end{aligned}
\tag{15.19}
$$

In deriving (15.19) we used the initial condition $\varrho(0) = |1\rangle\langle 1|$, so that

$$
\langle A_{12}(t_1)\, A_{21}(t_2)\rangle = \exp\{[i\omega_{12} - (\gamma_{21} + \gamma_{32})]\,(t_1 - t_2) - 2\gamma_{21}t_2\}, \tag{15.20a}
$$

$$
\begin{aligned}
\langle A_{23}(t_1)\, A_{32}(t_2)\rangle &= \exp\{[i\omega_{23} - \gamma_{32}]\,(t_1 - t_2)\} \\
&\quad \cdot \gamma_{21}(\gamma_{21} - \gamma_{32})^{-1}\,[e^{-2\gamma_{32}t_2} - e^{-2\gamma_{21}t_2}].
\end{aligned}
\tag{15.20b}
$$

For some experimental implications of (15.19) we refer to [91]. To calculate the photon distribution we employ the analog of (7.23). Since we have assumed that the energy spectrum is nonequidistant, it is possible to associate a photon uniquely with each transition. The distribution of the photon emitted in the transition $|1\rangle \to |2\rangle$ will be given by

$$
p_{ks}^{12}(\infty) = |g_{ks}^{12}|^2 \int_0^\infty dt_1 \int_0^{t_1} dt_2\,\langle A_{12}(t_1)\, A_{21}(t_1)\rangle\, e^{-i\omega_{ks}(t_1 - t_2)} + \text{c.c.} \tag{15.21}
$$

Similarly the distribution of the photon emitted in the transition $|2\rangle \to |3\rangle$ will· be

$$
p_{ks}^{23}(\infty) = |g_{ks}^{23}|^2 \int_0^\infty dt_1 \int_0^{t_1} dt_2\,\langle A_{23}(t_1)\, A_{32}(t_2)\rangle\, e^{-i\omega_{ks}(t_1 - t_2)} + \text{c.c.}. \tag{15.22}
$$

On substituting (15.20) in (15.21) and (15.22) we obtain

$$
p_{ks}^{12}(\infty) = |g_{ks}^{12}|^2\,(\gamma_{21} + \gamma_{32})\,\gamma_{21}^{-1}\,[(\gamma_{21} + \gamma_{32})^2 + (\omega_{ks} - \omega_{12})^2]^{-1}, \tag{15.23}
$$

$$
p_{ks}^{23}(\infty) = |g_{ks}^{23}|^2\,\{\gamma_{32}^2 + (\omega_{ks} - \omega_{23})^2\}^{-1}, \tag{15.24}
$$

which are identical to the results obtained in § 3 by the Weisskopf-Wigner method and in Chapter 5 by the Goldberger-Watson method. In the above equations the renormalized frequencies appear, as does the Lamb shift of the ground state, which was missing from (5.34); we have thus removed the asymmetry occurring in Chapter 4 as well as in Chapter 5. The level shift of the ground state appears in (15.23) and (15.24) because we made the RWA on the master equation rather than on the Hamiltonian itself.

    In the above we have assumed that the energy spectrum of the atom was nonequidistant and nondegenerate. The master-equation is easily obtained even in cases when the spectrum is not of this type. To illustrate the method, we consider the cases of a three-level atom with equidistant or degenerate spectrum.

A) Three-Level Atom with Equidistant Spectrum

We assume that the only allowed transitions are $|1\rangle \to |2\rangle$, $|2\rangle \to |3\rangle$ and that ω is the energy separation between levels. We rewrite the interaction Hamiltonian in the form

$$H_1 = v_{12} A_{12} + v_{21} A_{21} + v_{13} A_{13} + v_{31} A_{31}. \tag{15.25}$$

The master equation for the reduced density operator (in Born, Markov and RWA) is found to be

$$\partial \varrho_{11}/\partial t = -2\gamma_1 \varrho_{11},$$

$$\partial \varrho_{22}/\partial t = 2\gamma_1 \varrho_{11} - 2\gamma_2 \varrho_{22},$$

$$\varrho_{33} = 1 - \varrho_{11} - \varrho_{22},$$

$$\partial \varrho_{13}/\partial t = -\gamma_1 \varrho_{13} - 2i\omega \varrho_{13}, \tag{15.26}$$

$$\partial \varrho_{12}/\partial t = -(\gamma_1 + \gamma_2) \varrho_{12} - i\omega' \varrho_{12},$$

$$\partial \varrho_{23}/\partial t = 2\gamma_0 \varrho_{12} - i\omega'' \varrho_{23} - \gamma_2 \varrho_{23},$$

where $\gamma_1, \gamma_2, \gamma_0$ are the damping coefficients

$$\gamma_1 = \tfrac{2}{3}(\omega/c)^3 |d_{12}|^2, \quad \gamma_2 = \tfrac{2}{3}(\omega/c)^3 |d_{23}|^2, \quad \gamma_0 = \tfrac{2}{3}(\omega/c)^3 \, d_{12} \cdot d_{23}, \tag{15.27a}$$

and ω' and ω'' are the effective transition frequencies

$$\omega' = \omega + \tfrac{2}{3}|d_{12}|^2 \pi^{-1} \int k^3 \, dk \, [(k + k_0)^{-1} - (k - k_0)^{-1}], \tag{15.27b}$$

$$\omega'' = \omega + \tfrac{2}{3}|d_{23}|^2 \pi^{-1} \int k^3 \, dk \, [(k + k_0)^{-1} - (k - k_0)^{-1}], \tag{15.27c}$$

which are to be renormalized. The equations for the diagonal elements are identical to (15.15). The off-diagonal elements behave slightly differently. One should also note a new damping coefficient γ_0 in the equidistant case, which appears only in the equations of motion for off-diagonal matrix elements. In the steady state $\varrho_{ij} = \delta_{i3}\delta_{j3}$, as expected. In the special case when $d_{12} = d_{23}$, the three-level (equidistant) atom problem is similar to that of a spin-1 system. One can show that the operators defined by

$$S^+ = \sqrt{2}\,(A_{12} + A_{23}), \quad S^- = \sqrt{2}\,(A_{21} + A_{32}), \quad S^z = (A_{11} - A_{33}), \tag{15.28}$$

satisfy the angular momentum commutation relations corresponding to spin-1 value, i.e.

$$S^2 = 2, \quad S^z |1\rangle = |1\rangle, \quad S^z |2\rangle = 0, \quad S^z |3\rangle = -|3\rangle. \tag{15.29}$$

B) Three-Level Atom with Degenerate Spectrum

Consider now a three-level atom whose levels $|1\rangle$ and $|2\rangle$ are degenerate so that the possible transitions are $|1\rangle \rightarrow |3\rangle$, $|2\rangle \rightarrow |3\rangle$. The matrix elements of the density operator are found to satisfy equations (ignoring the Lamb shift terms)

$$\partial \varrho_{11}/\partial t = -\kappa \{ 2\mu_1 \varrho_{11} + \mu \varrho_{21} + \mu^* \varrho_{12} \}, \tag{15.30a}$$

$$\partial \varrho_{22}/\partial t = -\kappa \{ 2\mu_2 \varrho_{22} + \mu \varrho_{21} + \mu^* \varrho_{12} \}, \tag{15.30b}$$

$$\partial \varrho_{21}/\partial t = -\kappa (\mu_1 + \mu_2) \varrho_{21} - \kappa \mu^* (\varrho_{11} + \varrho_{22}), \tag{15.30c}$$

$$\partial \varrho_{31}/\partial t = -\kappa \{ \mu_1 \varrho_{31} + \mu^* \varrho_{32} \},$$
$$\partial \varrho_{32}/\partial t = -\kappa \{ \mu_2 \varrho_{32} + \mu \varrho_{31} \}, \tag{15.30d}$$

where

$$\kappa = \tfrac{2}{3}(\omega/c)^3, \quad \mu_1 = |d_{13}|^2, \quad \mu_2 = |d_{23}|^2, \quad \mu = d_{13} \cdot d_{23}^*. \tag{15.30e}$$

Equation (15.30) are linear equations and can be solved easily. We consider here only the simplified situation when $\mu = \mu_1 = \mu_2$ and then, defining $\gamma = \kappa \mu$, we obtain the equations

$$\partial \varrho_{11}/\partial t = -\gamma (2\varrho_{11} + \varrho_{12} + \varrho_{21}),$$

$$\partial \varrho_{22}/\partial t = -\gamma (2\varrho_{22} + \varrho_{21} + \varrho_{12}),$$

$$\partial \varrho_{31}/\partial t = -\gamma (\varrho_{31} + \varrho_{32}) \tag{15.31}$$

$$\partial \varrho_{32}/\partial t = -\gamma (\varrho_{32} + \varrho_{31}),$$

$$\partial \varrho_{21}/\partial t = -2\gamma \varrho_{21} - \gamma (\varrho_{11} + \varrho_{22}),$$

which admit a constant of integration

$$\varrho_{11} + \varrho_{22} - \varrho_{12} - \varrho_{21} = \alpha, \tag{15.32}$$

and then

$$\varrho_{11}(t) + \varrho_{22}(t) = e^{-4\gamma t} (\varrho_{11}(0) + \varrho_{22}(0)) + \tfrac{1}{2}\alpha (1 - e^{-4\gamma t}),$$

$$\varrho_{11}(t) - \varrho_{22}(t) = e^{-2\gamma t} (\varrho_{11}(0) - \varrho_{22}(0)),$$

$$\varrho_{21}(t) - \varrho_{12}(t) = e^{-2\gamma t} (\varrho_{21}(0) - \varrho_{12}(0)).$$

It should be noted that the above results for the degenerate case are very different from those for the nondegenerate case. The equations of motion for the off-diagonal elements and diagonal elements are mixed. For the initial excitation

$$\varrho(0) = |1\rangle \langle 1|, \tag{15.33}$$

we have

$$\varrho_{11}(t) = \tfrac{1}{4}(1 + e^{-4\gamma t}) + \tfrac{1}{2} e^{-2\gamma t},$$
$$\varrho_{22}(t) = \tfrac{1}{4}(1 + e^{-4\gamma t}) - \tfrac{1}{2} e^{-2\gamma t},$$
$$\varrho_{12}(t) = \varrho_{21}(t) = -\tfrac{1}{4}(1 - e^{-4\gamma t}),$$

implying that in the steady state

$$\varrho_{11}(\infty) = \varrho_{22}(\infty) = \tfrac{1}{4},$$
$$\varrho_{33}(\infty) = \tfrac{1}{2}, \tag{15.34}$$
$$\varrho_{12}(\infty) = \varrho_{21}(\infty) = -\tfrac{1}{4}.$$

This behavior (15.34) may seem somewhat surprising because one would expect that in the steady state the atom would remain in the ground state. On the other hand, for the initially symmetric excitation

$$\varrho(0) = |\psi\rangle \langle \psi|, \quad |\psi\rangle = 2^{-\tfrac{1}{2}}(|1\rangle + |2\rangle),$$
$$\varrho_{11}(0) = \varrho_{22}(0) = \varrho_{12}(0) = \varrho_{21}(0) = \tfrac{1}{2},$$

we find

$$\varrho_{11}(t) = \varrho_{22}(t) = \varrho_{12}(t) = \varrho_{21}(t) = \tfrac{1}{2} e^{-4\gamma t},$$

and hence in the steady state

$$\varrho_{11} = \varrho_{22} = \varrho_{12} = \varrho_{21} = 0,$$
$$\varrho_{33} = 1, \tag{15.35}$$

and the atom is left in the ground state. The steady state behavior can also be discussed in rather general terms (see also Appendix C). For the case we are discussing the interaction with the radiation field is through a combination of the atomic operators:

$$S^+ = 2^{-\tfrac{1}{2}}(A_{13} + A_{23}), \quad S^- = 2^{-\tfrac{1}{2}}(A_{31} + A_{32}),$$
$$S^z = \tfrac{1}{4}\{A_{22} + A_{11} - 2A_{33} + A_{12} + A_{21}\}. \tag{15.36}$$

It can be shown that these operators satisfy the same commutation relations as the spin-$\tfrac{1}{2}$ operators, i.e.

$$S^+ S^+ = S^- S^- = 0, \quad S^+ S^z = -S^z S^+ = -\tfrac{1}{2} S^+, \quad S^z S^z = \tfrac{1}{4} \text{ etc.}$$

The eigenstates of S^z are given by

$$S^z|\pm\rangle = \pm \tfrac{1}{2}|\pm\rangle, \quad |+\rangle = 2^{-\frac{1}{2}}(|1\rangle + |2\rangle), \quad |-\rangle = |3\rangle. \tag{15.37}$$

Since the coupling in the interaction Hamiltonian is via the operators defined by (15.36), one would expect the system to be found in the ground state of S^z only if it is prepared at time $t = 0$ in a state which is the linear combination of the states $|\pm\rangle$ as defined by (15.37). From this we see again that the symmetric state will decay to the ground state $|3\rangle$, whereas the unsymmetrized state (15.33) could result in different behavior as it cannot be expressed as a linear combination of $|\pm\rangle$.

For completeness, we also give the Langevin equations describing spontaneous emission from a multilevel atom with nondegenerate and nonequidistant spectrum. These are obtained in the same way as in Chapter 8:

$$\dot{A}_{ij} = \delta_{ij} \sum_{k \neq j} 2\gamma_{jk} A_{kk} - \tfrac{1}{2}(\Gamma_i + \Gamma_j) A_{ij} + F_{ij}, \tag{15.38}$$

where the random-force operator F_{ij} has the property

$$\langle F_{ij} \rangle = 0, \quad \langle F_{ij}(t) F_{kl}(t') \rangle = 2 \langle D_{ijkl} \rangle \, \delta(t - t'), \tag{15.39}$$

where the diffusion coefficient is given by the Einstein relation

$$2 \langle D_{ijkl} \rangle = \delta_{jk} \delta_{il} \sum_{m \neq l} 2\gamma_{lm} \langle A_{mm} \rangle - 2\gamma_{jk} \delta_{ij} \langle A_{kl} \rangle$$
$$- \delta_{kl} 2 \langle A_{ij} \rangle \, \gamma_{1j} + \delta_{jk} \langle A_{il} \rangle \, \Gamma_j. \tag{15.40}$$

The higher-order correlations of the random force can be calculated using (8.3b).

Finally, we mention that concepts like radiation-reaction fields can also be discussed for the case of a multilevel atom. The discussion proceeds along the lines of Chapter 8, therefore we ignore it. The case of many multilevel atoms is extremely cumbersome, and we hope to discuss it elsewhere; for some aspects of it we refer to [92]. Here we simply quote the master equation describing the emission from a collection of identical multilevel atoms (with nonequidistant and nondegenerate spectrum)

$$\partial \varrho / \partial t = -i \sum_{j\alpha} [E_\alpha A_{\alpha\alpha}^{(j)}, \varrho] - i \sum_{i \neq j \beta > \alpha} \Omega_{\beta\alpha}^{ij} [A_{\alpha\beta}^{(j)} A_{\beta\alpha}^{(i)}, \varrho]$$
$$- \sum_{ij\beta > \alpha} \gamma_{\beta\alpha}^{ij} \{ \varrho A_{\alpha\beta}^{(j)} A_{\beta\alpha}^{(i)} - A_{\beta\alpha}^{(i)} \varrho A_{\alpha\beta}^{(j)} + \text{H.C.} \}, \tag{15.41}$$

where E_α are the renormalized energy levels, $A_{\alpha\beta}^{(j)}$ are defined by [cf. (2.38)]

$$A_{\alpha\beta}^{(j)} = |\alpha\rangle_j \, {}_j\langle\beta|,$$

and $\gamma_{\alpha\beta}^{ij}$ and $\Omega_{\alpha\beta}^{ij}$ are defined as before [cf. (6.45) and (6.52)]. The total radiation rate is now given by

$$I(t) = 2 \sum_{ij\beta>\alpha} \gamma_{\beta\alpha}^{ij} \omega_{\alpha\beta} \langle A_{\alpha\beta}^{(j)} A_{\beta\alpha}^{(i)} \rangle . \tag{15.42}$$

The analog of (13.9) for the case of a small sample of three-level atoms is

$$\partial p(n_1, n_2)/\partial t = 2\sum \gamma_{\alpha\beta} \{ q_\alpha(q_\beta + 1) \, p(n_1 + \mu_{\alpha\beta}, n_2 + \nu_{\alpha\beta})$$
$$- q_\beta(q_\alpha + 1) \, p(n_1, n_2) \}, \quad \mu_{31} = \mu_{21} = (1 - \mu_{32}) = 1, \tag{15.43}$$
$$\nu_{31} = \nu_{32} = (1 - \nu_{21}) = 1 .$$

Here $p(n_1, n_2)$ is the diagonal matrix element of ϱ with respect to the collective states $|N, n_1, n_2\rangle$ (analog of Dicke states). In the state $|N, n_1, n_2\rangle$ n_1 atoms are in the uppermost state $|1\rangle$, $(N - n_2)$ atoms in the ground state $|3\rangle$, and $(n_2 - n_1)$ atoms in the state $|2\rangle$. The terms $2\gamma_{31}(N - n_2 + 1)n_1$, $2\gamma_{32}(n_2 - n_1)(N - n_2 + 1)$ and $2\gamma_{21}(n_2 - n_1 + 1)n_1$ are the respective transition probabilities per unit time that the system will make a transition from the state $|N, n_1, n_2\rangle$ to $|N, n_1 - 1, n_2 - 1\rangle$, $|N, n_1, n_2 - 1\rangle$, $|N, n_1 - 1, n_2\rangle$.

16. Neoclassical Theory of Spontaneous Emission

In this section we review the neoclassical theory of emission due to Jaynes and coworkers [9–12]. However, this theory differs in its predictions from that of quantum electrodynamics and recently some doubts have been expressed as to its correctness. We also discuss at each stage the relation [22] between the quantum electrodynamic (QED) results and the neoclassical (NC) results and point out how the transition from QED to NC can be "formally" made.

Consider a two-level atom with energy levels E_1 and E_2 with wave functions ψ_i. The wave function of the atom at time t is a linear superposition of ψ_1 and ψ_2, i.e.

$$\psi(t) = \alpha(t) \, \psi_1 + \beta(t) \, \psi_2 . \tag{16.1}$$

The interaction between the electromagnetic field and the atom is taken to be $- d \cdot E$. The field E in the neoclassical theory is a c-number field. We will specify it later. The Schrödinger equation leads (we follow closely Ref. [10]) to

$$i\dot{\alpha} = E_1\alpha - d \cdot E\beta ,$$
$$i\dot{\beta} = E_2\beta - d \cdot E\alpha , \tag{16.2}$$

---

[22] For some comments on neoclassical theory, see also [93].

where d is now the matrix element, which is assumed to be real, between the states ψ_1 and ψ_2. We introduce the bilinear forms defined by

$$s^+ = \alpha^* \beta, \quad s^- = \alpha \beta^*, \quad s^z = \tfrac{1}{2}(\alpha \alpha^* - \beta \beta^*), \tag{16.3}$$

then the equations of motion for $s^{\pm}, s^z$ are

$$\dot{s}^{\pm} = \pm i \omega s^{\pm} \pm 2i(d \cdot E) s^z,$$
$$\dot{s}^z = i(d \cdot E)(s^+ - s^-). \tag{16.4}$$

It should be noted that the energy of the atom $W(t)$ and its dipole moment are given by

$$W(t) = \omega s^z, \quad M(t) = (s^+ + s^-). \tag{16.5}$$

From (16.4), (16.5) it is evident that the energy and dipole moment satisfy the equations

$$\ddot{M} + \omega^2 M = -4(d \cdot E) W,$$
$$\dot{W} = \dot{M}(d \cdot E). \tag{16.6}$$

The first of the Eqs. (16.6) has a clear physical meaning: it is the equation of motion of a driven harmonic oscillator, the driving force itself is, however, proportional to the energy of the atom and therefore the response of the dipole depends on the energy and is in phase (out of phase) when W is positive (negative). This equation admits a constant of motion, viz.

$$\dot{M}^2 + \omega^2 M^2 + 4W^2 = \text{constant} = 4\omega^2 \{s^+ s^- + s^z s^z\}$$
$$= \omega^2 (|\alpha|^2 + |\beta|^2)^2 = \omega^2, \tag{16.7}$$

where we use the normalization condition $|\alpha|^2 + |\beta|^2 = 1$. We now introduce polar coordinates defined by

$$s^z = -\tfrac{1}{2}\cos\theta, \quad s^{\pm} = \tfrac{1}{2}\sin\theta \, e^{\pm i\varphi},$$

and then we have   (16.8)

$$\dot{M} \pm i\omega M = \omega \sin\theta \, e^{\pm i\varphi}.$$

We have so far *not* specified the electric field which appears in the above equations. According to this theory the electric field $E(t)$ is given by

$$E(t) = E_{\text{ext}}(t) + E_{RR}(t), \tag{16.9}$$

where E_{ext} is the applied (external field) and E_{RR} is the "radiation-reaction" field. In what follows we will assume that there is *no* applied field. We take for the radiation reaction the value obtained by Stroud and Jaynes:

$$E_{RR}(t) = \tfrac{2}{3}(d/c^3)\,\dddot{M}(t) - \tfrac{4}{3}(K/\pi c^3)\,d\,\ddot{M}(t),\tag{16.10}$$

where K is the cutoff frequency. The reaction field (16.10) can be approximated by

$$E_{RR}(t) = -\tfrac{2}{3}(\omega^2/c^3)\,d\dot{M} + \tfrac{4}{3}(K/\pi c^3)\,d\,\omega^2 M,\tag{16.11}$$

since the dipole moment oscillates at roughly frequency ω. Let us introduce the parameters γ, Ω_{NC} defined by

$$\gamma = \tfrac{2}{3}(\omega/c)^3|d|^2,\qquad \Omega_{NC} = -2\gamma K/\pi\omega = -\tfrac{4}{3}(K/\pi c^3)|d|^2\omega^2.\tag{16.12}$$

We will see later that γ is related to the damping and Ω_{NC} to the Lamb shift. On substituting (16.10) in (16.4) and on transforming to the "interaction picture" we obtain the equations of motion

$$\dot{s}^{\pm,I} = \mp 2i\Omega_{NC}(s^{+I}e^{i\omega t} + s^{-I}e^{-i\omega t})\,s^z\,e^{\mp i\omega t}$$

$$\mp (2i\gamma/\omega)\,e^{\mp i\omega t}\{\dot{s}^{+I}e^{i\omega t} + \dot{s}^{-I}e^{-i\omega t} + i\omega s^{+I}e^{i\omega t}$$

$$- i\omega s^{-I}e^{-i\omega t}\}\,s^z,\tag{16.13}$$

$$\dot{s}^z = -i(s^{+I}e^{i\omega t} - s^{-I}e^{-i\omega t})\,[\Omega_{NC}(s^{+I}e^{i\omega t} + s^{-I}e^{-i\omega t})$$

$$+ (\gamma/\omega)(\dot{s}^{+I}e^{i\omega t} + \dot{s}^{-I}e^{-i\omega t} + i\omega s^{+I}e^{i\omega t} - i\omega s^{-I}e^{-i\omega t})],$$

where $s^{\pm I}$ are the slowly varying quantities defined by

$$s^{\pm,I} = s^{\pm}\,e^{\mp i\omega t}.\tag{16.14}$$

We now make the "rotating-wave approximation", i.e. ignore the rapidly oscillating terms from (16.13). We then obtain

$$\dot{s}^{\pm I} = (2\gamma s^{\pm I} \mp 2i\Omega_{NC}s^{\pm I})\,s^z,$$

$$\dot{s}^z = -2\gamma s^{+I}s^{-I},$$

which on transforming back to the Schrödinger picture reduce to

$$\dot{s}^{\pm} = \pm i\omega s^{\pm} + 2\gamma s^{\pm}s^z \mp 2i\Omega_{NC}s^{\pm}s^z,$$

$$\dot{s}^z = -2\gamma s^{+}s^{-}.\tag{16.15}$$

Equations (16.15) are the basic equations of neoclassical theory describing the spontaneous emission from a two-level atom in the dipole approximation, RWA, and the approximation leading from (16.10) to (16.11).

We first look at the solutions of (16.15) and compare them with the results of QED. From (16.15) we have

$$s^{\pm}(t) = s^{\pm}(0) \exp\left\{\pm i\omega t + (2\gamma \mp 2i\Omega_{NC}) \int_0^t s^z(t')\, dt'\right\}. \tag{16.16a}$$

The second of Eqs. (16.15) can be written as

$$\dot{s}^z = -2\gamma/4 + 2\gamma s^z s^z,$$

the solution of which is

$$s^z(t) = -\tfrac{1}{2}\tanh\gamma(t - t_0). \tag{16.16b}$$

The constant of integration t_0 is determined from the initial condition. If the atom is in the upper state at time $t = 0$, $s^z(0) = \tfrac{1}{2}$, then (16.16b) implies that the atom *remains for ever* in its upper state. In the neoclassical theory this is the point of unstable equilibrium and even a slight perturbation will cause the system to decay. The dipole moment is given by

$$s^{\pm}(t) = s^{\pm}(0)\, e^{\pm i\omega t} \left\{\frac{\cosh\gamma(t - t_0)}{\cosh\gamma t_0}\right\}^{-(1 \mp i\Omega_{NC}/\gamma)}. \tag{16.16c}$$

From (16.16a) it is clear that the frequency of the atom becomes *time-modulated* as a result of spontaneous emission; the modulation depends on the energy and the time-dependent frequency is $\omega - \Omega_{NC}(\omega + \Omega_{NC})$ when the atom is near its upper state (lower state). The long-time behavior of the energy is given by

$$s^z \to -\tfrac{1}{2}(1 - 2e^{-2\gamma t}), \quad t \to \infty. \tag{16.17}$$

The intensity of the emitted radiation $I(t)$ is equal to

$$I(t) = -\omega(\partial/\partial t)\, s^z = \tfrac{1}{2}\gamma\omega\, \mathrm{Sech}^2\,\gamma(t - t_0). \tag{16.18}$$

The line shape of the emitted radiation is proportional to $|s^+(\omega) + s^-(\omega)|^2$ which is given by

$$p(\omega_{\mathrm{rad}}) = (\pi/\Omega_{NC}\gamma)\sinh(\Omega_{NC}/\gamma)\left[\cosh(\Omega_{NC}/\gamma) + \cosh(\omega_{\mathrm{rad}} - \omega)/\gamma\right]^{-1}. \tag{16.19}$$

The results given by (16.16), (16.18) and (16.19) are very different from those predicted by QED, which are

$$\begin{aligned} s^{\pm}(t) &= s^{\pm}(0)\exp\{\pm(i\omega + \Omega_{ii})\,t - \gamma t\},\\ s^z(t) &= -\tfrac{1}{2} + (\tfrac{1}{2} + s^z(0))\, e^{-2\gamma t}, \end{aligned} \tag{16.20}$$

and the line shape of the emitted radiation is Lorentzian. We also note that there is *no* time-dependent frequency modulation. The decay constant γ is identical in both theories whereas Ω_{ii} is very different from Ω_{NC} in the neoclassical theory. Let us examine the reasons for some of these discrepancies. The master equation describing the spontaneous emission is given by [cf. Eq. (10.1)]

$$\partial\varrho/\partial t = -i\omega[S^z,\varrho] - \gamma(S^+S^-\varrho - 2S^-\varrho S^+ + \varrho S^+S^-)$$
$$- \tfrac{1}{2}i(\Omega_{ii}+\Omega_{ii}^-)[S^+S^-,\varrho] + \tfrac{1}{2}i(\Omega_{ii}-\Omega_{ii}^-)[S^-S^+,\varrho], \qquad (16.21)$$

where $\Omega_{ii}, \Omega_{ii}^-$ are given by Eqs. (6.48) and (6.66), respectively. From (16.21) we see that the mean value of an operator Q obeys the equation

$$\partial\langle Q\rangle/\partial t = i\omega\langle[S^z,Q]\rangle + \tfrac{1}{2}(\Omega_{ii}+\Omega_{ii}^-)\langle[S^+S^-,Q]\rangle$$
$$- \tfrac{1}{2}(\Omega_{ii}-\Omega_{ii}^-)\langle[S^-S^+,Q]\rangle - \gamma\langle[Q,S^+]S^- + S^+[S^-,Q]\rangle, \qquad (16.22)$$

and in particular we obtain

$$(\partial/\partial t)\langle S^+\rangle = +i\omega\langle S^+\rangle - i(\Omega_{ii}+\Omega_{ii}^-)\langle S^+S^z\rangle + i(\Omega_{ii}-\Omega_{ii}^-)$$
$$\cdot\langle S^zS^+\rangle + 2\gamma\langle S^+S^z\rangle, \qquad (16.23\text{a})$$

$$(\partial/\partial t)\langle S^z\rangle = -2\gamma\langle S^+S^-\rangle. \qquad (16.23\text{b})$$

If we *overlook* the *fact that* $S^\pm, S^z$ refer to the spin angular momentum operators corresponding to spin-$\tfrac{1}{2}$ value and use a semiclassical approximation in which the mean value of the product of two operators is replaced by the product of the mean values, then (16.23) reduce to

$$(\partial/\partial t)s^\pm = \pm i\omega s^\pm \mp 2i\Omega_{ii}^- s^z s^+ + 2\gamma s^\pm s^z,$$
$$(\partial/\partial t)s^z = -2\gamma s^+ s^-. \qquad (16.24)$$

Equation (16.24) are precisely the equations of the neoclassical theory and the term Ω_{ii}^- is also approximately equal to Ω_{NC} in the neoclassical equations. If, on the other hand, we take proper account of the anti-commutation relations, we get from (16.23) the quantum electrodynamic equations

$$(\partial/\partial t)\langle S^\pm\rangle = \pm i(\omega+\Omega_{ii})\langle S^\pm\rangle - \gamma\langle S^\pm\rangle,$$
$$(\partial/\partial t)\langle S^z\rangle = -2\gamma\langle S^z + \tfrac{1}{2}\rangle, \qquad (16.25)$$

which are linear equations. The linearity is due to the spin-$\tfrac{1}{2}$ nature of $S^\pm, S^z$. In the QED case the analog of (16.7) is the relation

$$\langle S\cdot S\rangle = \tfrac{3}{4}. \qquad (16.26)$$

Further, in the QED case the atomic-density operator has the property (10.7), viz.

$$\varrho^2(t) \neq \varrho(t),$$

i.e. it does not represent *a pure state*. The neoclassical theory right from the start assumes a pure state where it turns out that $\alpha(t)$ and $\beta(t)$ have well-defined values. For the case when the system starts in the upper state QED gives [cf. (10.6)]

$$\varrho(t) = \tfrac{1}{2} + 2\langle S^z(t)\rangle\, S^z = \tfrac{1}{2} + (2\,e^{-2\gamma t} - 1)\,S^z$$
$$= \overline{(\alpha(t)\,|1\rangle + \beta(t)\,|2\rangle)(\alpha^*(t)\,\langle 1| + \beta^*(t)\,\langle 2|)}$$

where the bar denotes the ensemble average with respect to the relative phase of α and β. This relative phase is "completely random". It should be further noted that in the neoclassical theory the electric field is treated as a classical quantity though the source of this electric field is the atomic system, which is treated quantum-mechanically. This is certainly an inconsistency in itself. Moreover, the appearance of nonlinear equations for the dipole moment etc. can also be traced back if we ignore all the "correlations" between the atomic system and the radiation field [cf. Eq. (8.53) of the mean-field theory].

For the harmonic oscillator model the QED and NC equations for the mean value of the dipole moment are same. In place of (16.6) one has

$$\ddot{M} + \omega^2 M \approx \tfrac{2}{3}(|d|/c^3)\,\omega^2\,\dot{M} - \tfrac{4}{3}(K/\pi c^3)\,|d|\,\omega^2 M, \tag{16.27}$$

which in RWA leads to the same equations as QED, viz.

$$\langle \dot{a}\rangle = -i\omega_0\langle a\rangle - \gamma\langle a\rangle, \tag{16.28}$$

where the renormalized frequency ω_0 has approximately the same value as in QED.

We have so far considered the spontaneous emission from a system of one two-level atom. Similar results hold for an atom with many levels [11]. We discuss only briefly the case of a multilevel atom with a nondegenerate and nonequidistant spectrum. As before, we denote the energy eigenvalues by E_i and the eigenfunctions by ψ_i. The wave function at time t is written as

$$\psi(t) = \Sigma\,\alpha_i(t)\,\psi_i. \tag{16.29}$$

We introduce the quadratic forms defined by

$$c_{ij} = \alpha_j^*\,\alpha_i,$$

these satisfy

$$\dot{c}_{ij} = -i\omega_{ij}c_{ij} + i\sum_m (d_{im}\cdot E_{RR}c_{mj} - c_{im}d_{mj}\cdot E_{RR}), \tag{16.30}$$

where d_{ij} are the dipole moment matrix elements between the states i and j. The radiation-reaction field is written as

$$E_{RR} = \sum_{pq} c_{pq} \beta_{pq} \,. \tag{16.31}$$

On substituting (16.31) in (16.30) and ignoring the rapidly oscillating terms, we obtain

$$\dot{c}_{ij} = -i\omega_{ij} c_{ij} + i \sum_q c_{qq} (d_{iq} \cdot \beta_{iq} - d_{qj} \cdot \beta_{qj}) c_{ij}$$
$$= -i \left[\omega_{ij} - \sum_q (\Omega_{iq} - \Omega_{qj}) c_{qq} \right] c_{ij} - \sum_q (\gamma_{iq} + \gamma_{jq}) c_{qq} c_{ij}, \tag{16.32}$$

where Ω_{ij}, γ_{ij} now are different from (6.45) and (6.52). Here Ω's represent the frequency shifts and γ_{iq} are given by

$$\gamma_{iq} = \tfrac{2}{3} |d_{iq}|^2 (\omega_{iq}/c)^3 \,. \tag{16.33}$$

The structure of Eqs. (16.32) is very different from the QED equations of Chapter 15. This was excepted since the NC theory ignores all correlations between field and matter. The above equations again predict the time-dependent frequency modulation.

In particular, for a three-level atom with transitions allowed only between adjacent levels, we find that

$$\dot{c}_{11}/c_{11} = -2\gamma_{12} c_{22},$$
$$\dot{c}_{22}/c_{22} = 2\gamma_{12} c_{11} - 2\gamma_{23} c_{33}, \tag{16.34}$$
$$\dot{c}_{33}/c_{33} = 2\gamma_{23} c_{22},$$

which have the constant of integration

$$(c_{11})^{\gamma_{12}} (c_{33})^{\gamma_{23}} = \text{constant} \,. \tag{16.35}$$

The relation (16.35) implies, as in the case of a two-level atom, that the state $|1\rangle$ is a metastable state. The Eqs. (16.32) can be obtained from the master equation if the following property of the atomic operators

$$A_{ij} A_{kl} = A_{il} \delta_{jk}$$

is *not* used and a semiclassical approximation is made.

The inclusion of the applied external field is straightforward in NC theory and we refer to the original literature where explicit results pertaining to this case can be found. For example, the details of the NC theory of a small sample of a system of two-level atoms can be found in Ref. [86].

Finally, we mention how it is possible to derive the results of QED by modifying NC theory in the following manner:

(1) One regards the field $E(t)$ as a quantum-mechanical field which has as its source the oscillating dipole moments, i.e. the current operator is taken to be proportional to the dipole-moment operator.

(2) In the mean-value equations terms like $\langle(d \cdot E)S^z\rangle$ appear. E is now an operator which would be a linear combination of the operators $S^\pm, S^z$, and we put these products of operators in the normal order, i.e. we arrange all the positive-frequency part of the field operator to the right of the atomic operators and all the negative-frequency part of the field operator to the left of the atomic operators.

It can be shown that such a prescription leads to the QED Eqs. (16.25). We are, of course, motivated to adopt this kind of prescription by our QED calculations. To make this point very clear, we have from Eq. (7.20)

$$a_{ks}(t) = a_{ks}(0)\, e^{-i\omega_{ks}t} - ig_{ks}^* \int_0^t [S^+(\tau) + S^-(\tau)]\, e^{-i\omega_{ks}(t-\tau)}\, d\tau\,, \qquad (16.36)$$

and

$$-d \cdot E(t) = \sum_{ks} (g_{ks} a_{ks} + g_{ks}^* a_{ks}^+)$$

$$= \sum_{ks} (g_{ks} a_{ks}(0)\, e^{-i\omega_{ks}t} + \text{H.C.}) - i \sum_{ks} |g_{ks}|^2 \int_0^t [S^+(\tau) + S^-(\tau)]$$

$$\cdot e^{-i\omega_{ks}(t-\tau)}\, d\tau + i \sum_{ks} |g_{ks}|^2 \int_0^t [S^+(\tau) + S^-(\tau)]\, e^{i\omega_{ks}(t-\tau)}\, d\tau$$

$$= -d \cdot E_0^{(+)} - d \cdot E_0^{(-)} - d \cdot E_{RR}^{(+)} - d \cdot E_{RR}^{(-)}\,,$$

where

$$-d \cdot E_{RR}^{(+)} = -i \sum_{ks} |g_{ks}|^2 \int_0^t [S^+(\tau) + S^-(\tau)]\, e^{-i\omega_{ks}(t-\tau)}\, d\tau\,,$$

$$\qquad (16.37)$$

$$-d \cdot E_{RR}^{(-)} = +i \sum_{ks} |g_{ks}|^2 \int_0^t [S^+(\tau) + S^-(\tau)]\, e^{i\omega_{ks}(t-\tau)}\, d\tau\,.$$

Now the mean-value equation (16.4) will be written as

$$\langle \dot{S}^\pm \rangle = \pm i\omega \langle S^\pm \rangle \pm 2i \langle S^z (d \cdot E_{RR}^+) \rangle \pm 2i \langle (d \cdot E_{RR}^{(-)}) S^z \rangle\,,$$

$$\qquad (16.38)$$

$$\langle \dot{S}^z \rangle = i \langle d \cdot E_{RR}^{(-)}(S^+ - S^-) \rangle + i \langle (S^+ - S^-)\, d \cdot E_{RR}^{(+)} \rangle\,.$$

The Eqs. (16.38) lead to the QED equations if the conventional approximations (Born and Markov) are made in which the "radiation-reaction field" become

$$-d \cdot E_{RR}^+ = \tfrac{1}{2}(\Omega_{ii} + \Omega_{ii}^-)\, S^- - \tfrac{1}{2}(\Omega_{ii} - \Omega_{ii}^-)\, S^+ - i\gamma S^-\,,$$

$$\qquad (16.39)$$

$$-d \cdot E_{RR}^- = \tfrac{1}{2}(\Omega_{ii} + \Omega_{ii}^-)\, S^+ - \tfrac{1}{2}(\Omega_{ii} - \Omega_{ii}^-)\, S^- + i\gamma S^+\,,$$

so that the total reaction field is given by

$$-d \cdot E_{RR} = -i\gamma(S^- - S^+) + \Omega_{ii}^-(S^+ + S^-), \tag{16.40}$$

which compares very well with (16.11). The positive- and negative-frequency parts are given by (16.39), and it is because of our ordering prescription that Ω_{ii} appears in the equations for $\langle S^\pm \rangle$. It should be noted that the radiation reaction fields as given by (16.37) can be evaluated exactly. The case of many two-level atoms emitting spontaneously can be discussed similarly. We have in place of (16.38) the equations for each atom

$$\langle \dot{S}_i^+ \rangle = i\omega \langle S_i^+ \rangle + 2i \langle S_i^z(d \cdot E_{RR}^+) \rangle + 2i \langle (d \cdot E_{RR}^-) S_i^z \rangle$$
$$+ 2i \sum_{j \neq i} \langle S_i^z(d \cdot E_{ij}^+) \rangle + 2i \sum_{j \neq i} \langle (d \cdot E_{ij}^-) S_i^z \rangle, \tag{16.41}$$

$$\langle \dot{S}_i^z \rangle = \langle i(d \cdot E_{RR}^-)(S_i^+ - S_i^-) \rangle + \sum_{j \neq i} i \langle (d \cdot E_{ij}^{(-)})(S_i^+ - S_i^-) \rangle + \text{H.C.},$$

where E_{ij}^+ are the dipole fields which we have already evaluated [Eq. (8.50)]. On substituting (8.50) in (16.41) and making the rotating-wave approximation we obtain Eqs. (8.20) and (8.21).

17. Spontaneous Emission in Presence of a Thermal Field

As a further application of the master equation techniques, we consider spontaneous emission from an atom in presence of a thermal (black-body) field at a temperature T. The interaction Hamiltonian is given by (2.12) but now the initial state of the field is

$$\varrho_R(0) = \exp\left\{-\beta \sum_{ks} \omega_{ks} a_{ks}^+ a_{ks}\right\} \Big/ \text{Tr} \exp\left\{-\beta \sum_{ks} \omega_{ks} a_{ks}^+ a_{ks}\right\}, \tag{17.1}$$

where

$$\beta = 1/KT, \quad K = \text{Boltzmann constant}.$$

The mean occupation number of the mode ks of the free field is equal to

$$\langle n_{ks} \rangle = \langle a_{ks}^+ a_{ks} \rangle = (\exp(\beta\omega_{ks}) - 1)^{-1}, \tag{17.2}$$

i.e. each mode has a finite occupation number. We use the same kind of approximations as in Chapter 6 and Chapter 15 and we find that the reduced-density operator obeys the equation

$$\partial \varrho_{ij}/\partial t = -i\omega_{ij}\varrho_{ij} - \Gamma_{ji}\varrho_{ij} + \delta_{ij} \sum_{k \neq i} 2\gamma_{ik}\varrho_{kk}, \tag{17.3}$$

for a multilevel atom having a nonequidistant and nondegenerate spectrum. This master equation is the same as (15.5) except that now the transition from a level $|i\rangle$ to $|j\rangle$ is also allowed if $E_i < E_j$. γ_{kl} and γ_{lk} are related by

$$\gamma_{kl} = \gamma_{lk} \exp(\beta\omega_{lk}) . \tag{17.4}$$

In particular, for a two-level atom if we use the relations

$$S^+ = |1\rangle\langle 2| = A_{12}, \quad S^- = |2\rangle\langle 1| = A_{21}, \quad S^z = \tfrac{1}{2}(A_{11} - A_{22}),$$

we obtain the master equation

$$\partial\varrho/\partial t = -i\omega_0[S^z, \varrho] - \gamma(1 + \langle n(\omega)\rangle)(S^+ S^- \varrho - 2S^- \varrho S^+ + \varrho S^+ S^-)$$

$$- \gamma\langle n(\omega)\rangle (S^- S^+ \varrho - 2S^+ \varrho S^- + \varrho S^- S^+), \tag{17.5}$$

where $\gamma_{21} = \gamma(1 + \langle n(\omega)\rangle)$, $\gamma_{12} = \gamma\langle n(\omega)\rangle$ and γ is equal to half the Einstein A coefficient. ω_0 is the renormalized frequency

$$\omega_0 = \omega + \Delta , \tag{17.6}$$

where

$$\Delta = (1/2i)\sum_l \int_0^\infty dt \{\langle v_{1l}(t) v_{l1}(0)\rangle e^{-i\omega_{l1}t} + \langle v_{2l}(0) v_{l2}(t)\rangle e^{i\omega_{l2}t} - \text{c.c.}\}$$

$$= (1/2i)\int_0^\infty dt \{\langle v_{12}(t) v_{21}(0)\rangle e^{+i\omega t} + \langle v_{21}(0) v_{12}(t)\rangle e^{i\omega t} - \text{c.c.}\} \tag{17.7}$$

$$= \sum_{ks} |g_{ks}|^2 (1 + 2\langle n_{ks}\rangle) \{(\omega - \omega_{ks})^{-1} + (\omega + \omega_{ks})^{-1}\}$$

$$= \Delta_0 + \Delta_T .$$

Here Δ_0 is the usual Lamb shift whose renormalized value is given by (6.48) and Δ_T is a temperature-dependent shift. The temperature-dependent shift of the ground state is the negative of the temperature-dependent shift of the excited state and hence the factor 2 appears in (17.7). On simplification Δ_T reduces to

$$\Delta_T = (4\gamma/\pi\beta^2\omega^2)\int_0^\infty \frac{y^3\,dy}{(e^y - 1)(\beta^2\omega^2 - y^2)} , \tag{17.8}$$

which is a well-defined quantity. The occurrence of y^3 in the integrand is due to the interaction Hamiltonian $-d \cdot E$ (rather than $-A \cdot p$ which was used by Walsch [94], see also [95]). It should be noted that for the computation of Δ_T, the term $2\pi \int |p|^2 d^3r$ does not contribute. At low temperatures Δ_T is proportional to T^4 ($-A \cdot p$ interaction leads to the wrong temperature dependence T^2 at low temperatures). For numerals concerning temperature-dependent shifts we refer to [94, 95].

It is clear from (17.5) that

$$(\partial/\partial t)\langle S^+ S^-\rangle = -2\gamma(1 + 2\langle n(\omega)\rangle)(\langle S^z\rangle - \langle S^z\rangle_{st}),$$

$$(\partial/\partial t)\langle S^\pm\rangle = [\pm i\omega_0 - \gamma(1 + 2\langle n(\omega)\rangle)]\langle S^\pm\rangle, \qquad (17.9)$$

$$\langle S^z\rangle_{st} = -\tfrac{1}{2}\tanh(\tfrac{1}{2}\beta\omega).$$

The effective decay constant is therefore

$$\gamma_T = \gamma(1 + 2\langle n(\omega)\rangle). \qquad (17.10)$$

The solution of (17.9) is

$$\langle S^z\rangle_t = \langle S^z\rangle_{st} + (\langle S^z\rangle_0 - \langle S^z\rangle_{st}) e^{-2\gamma_T t}, \qquad (17.11)$$

$$\langle S^\pm\rangle_t = \langle S^\pm\rangle_0 \exp\{\pm i\omega_0 t - \gamma_T t\}.$$

The rate at which the atom dissipates energy is

$$I(t) = -\omega(\partial/\partial t)\langle S^z\rangle = \gamma\omega\{1 + \coth\tfrac{1}{2}\beta\omega\} e^{-2\gamma_T t}, \qquad (17.12)$$

where we have assumed that the atom was in the excited state at time $t = 0$. The two-time correlation function is given by

$$\langle S^+(t) S^-(t')\rangle = \tfrac{1}{2}\exp\{(i\omega_0 - \gamma_T)(t - t')\}$$

$$\cdot [(1 - \tanh\tfrac{1}{2}\beta\omega) + (1 + \tanh\tfrac{1}{2}\beta\omega) e^{-2\gamma_T t'}]. \qquad (17.13)$$

Finally, it should be noted that the master equation (17.5) also describes spontaneous emission from a collection of identical two-level atoms confined to a region smaller than a wavelength (with frequency shifts ignored). $S^\pm, S^z$ are then the collective operators of Dicke. The rate at which the atom dissipates energy is now

$$(\partial/\partial t)\langle S^z\rangle = -2\gamma\langle S^+ S^-\rangle - 4\gamma\langle n(\omega)\rangle\langle S^z\rangle. \qquad (17.14)$$

Diagonal matrix elements in terms of Dicke states satisfy the Pauli master equation

$$\partial\varrho_{mm}/\partial t = 2\gamma(1 + \langle n(\omega)\rangle)(v_{m+1}\varrho_{m+1,m+1} - v_m\varrho_{m,m})$$

$$- 2\gamma\langle n(\omega)\rangle (v_{m+1}\varrho_{m,m} - v_m\varrho_{m-1,m-1}). \qquad (17.15)$$

The steady-state solution of (17.15), which can be obtained easily from microreversibility, is

$$\varrho_{mm} \propto e^{-\beta\omega m}. \qquad (17.16)$$

The master equation (17.15) can be handled in the same way as (13.5).

18. Spontaneous Emission in Presence of External Fields

In our treatment of spontaneous emission by means of master equations we have so far assumed that there was no coupling between different atomic levels, or that different two-level atoms were uncoupled except through the radiation field, or that no external fields were present. The analysis of the earlier section is easily extended to take such interactions into account. We first discuss the case of c-number (classical) external fields and write the interaction Hamiltonian in the form

$$H = H_A + H_R + H_{AR} + H_{ext}(t), \tag{18.1}$$

where $H_{ext}(t)$ is the interaction Hamiltonian between the atoms and the external field. On using the formalism of Chapter 6 one finds that the reduced-density operator in the Born and Markovian approximations satisfies the equation

$$(\partial/\partial t)\{\mathscr{P}\varrho(t)\} + i[H_{ext}(t), \mathscr{P}\varrho(t)] + \int_0^\infty d\tau\, \varrho_R(0) \tag{18.2}$$

$$\cdot \mathrm{Tr}_R\{[H_{AR}(t), [V(t, t-\tau) H_{AR}(t-\tau) V^+(t, t-\tau), \mathscr{P}\varrho(t)]]\}\} = 0,$$

where

$$V(t, \tau) = T \exp\left\{-i\int_\tau^t dt'\, H_{ext}(t')\right\}, \tag{18.3}$$

and all the operators are in the interaction picture. If the interaction between the external field and the atoms is not too strong (i.e. the Raabi frequency is much smaller than the atomic frequency), then one can ignore the evolution of the system under the influence of H_{ext} over the correlation time, i.e. one can put $V(t, t-\tau) \approx 1$. In such cases (18.2) simplifies to

$$(\partial/\partial t)\{\mathscr{P}\varrho(t)\} + i[H_{ext}(t), \mathscr{P}\varrho(t)]$$

$$+ \int_0^\infty d\tau\, \mathscr{P}[H_{AR}(t), [H_{AR}(t-\tau), \mathscr{P}\varrho(t)]] = 0, \tag{18.4}$$

which simply implies that one can superimpose the effects of the external field (coherent interaction) and vacuum fluctuations (incoherent interaction). A master equation of the form (18.2) has been used to discuss the "interference effects" between coherent and incoherent interactions as well as the dynamics of strongly interacting systems [96]. For the usual field strengths, the interference effects in the present context are negligible. We now apply (18.4) to some specific examples.

A) As a first example we consider the three-level atom (discussed in connection with the Goldberger-Watson type of approach) with levels

$|1\rangle$ and $|2\rangle$ coupled by H_{ext} and levels $|2\rangle$ and $|3\rangle$ radiatively coupled. The master equation according to (18.4) and (10.1) is

$$\partial\varrho/\partial t = -i\sum_i E_i[A_i^+ A_i, \varrho] - i[H_{\text{ext}}, \varrho] \tag{18.5}$$

$$-\gamma(A_{23}A_{32}\varrho - 2A_{32}\varrho A_{23} + \varrho A_{23}A_{32}),$$

with H_{ext} connecting only levels $|1\rangle$ and $|2\rangle$; 2γ is equal to the inverse lifetime of the state $|2\rangle$. In terms of the matrix elements we find

$$\partial\varrho_{11}/\partial t = -iX,$$

$$\partial\varrho_{22}/\partial t = iX - 2\gamma\varrho_{22},$$

$$\partial\varrho_{33}/\partial t = 2\gamma\varrho_{22}, \quad \partial X/\partial t \equiv (\partial/\partial t)(V_{12}\varrho_{21} - V_{21}\varrho_{12})$$

$$= -\gamma X - i\omega_{21}Y - 2i|V_{12}|^2(\varrho_{11} - \varrho_{22}), \tag{18.6}$$

$$\partial Y/\partial t \equiv (\partial/\partial t)(V_{12}\varrho_{21} + V_{21}\varrho_{12}) = -\gamma Y - i\omega_{21}X,$$

$$\partial\varrho_{32}/\partial t = -i\omega_{32}\varrho_{32} + iV_{12}\varrho_{31} - \gamma\varrho_{32},$$

$$\partial\varrho_{31}/\partial t = -i\omega_{31}\varrho_{31} + i\varrho_{32}V_{21}.$$

Equations (18.6) are linear equations and are easily solved. The solution for some matrix elements when the atom was initially in the state $|1\rangle$ is given by:

$$\varrho_{11}(t) = \tfrac{1}{4}(P^2 + Q^2)^{-1}\{[(\gamma^2 + P^2)(\gamma + Q)^2 e^{-(\gamma - Q)t} + Q \rightarrow -Q]$$

$$- [e^{iPt - \gamma t}(iP + \gamma)^2(Q^2 - \gamma^2) + P \rightarrow -P]\}, \tag{18.7}$$

$$\varrho_{22}(t) = \alpha(P^2 + Q^2)^{-1} e^{-\gamma t}[\cosh Qt - \cos Pt], \tag{18.8}$$

$$\varrho_{21}(t) = V_{21} e^{-\gamma t}(P^2 + Q^2)^{-1}\{(\omega - i\gamma)\cosh Qt$$

$$+ Q^{-1}\sinh Qt(\omega\gamma - \omega^2 - Q^2) - (\omega - i\gamma)\cos Pt \tag{18.9}$$

$$- P^{-1}\sin Pt(\omega\gamma - \omega^2 + P^2)\},$$

where

$$\alpha = 2|V_{21}|^2, \quad P + iQ = [(\omega_{12}^2 + \gamma^2)^2 + 4\alpha^2 + 4\alpha(\omega_{12}^2 - \gamma^2)]^{\tfrac{1}{4}}$$

$$\cdot \exp\left\{\tfrac{1}{2}i \tan^{-1}\left(\frac{-2\omega_{12}\gamma}{\omega_{12}^2 - \gamma^2 + 2\alpha}\right)\right\}. \tag{18.10}$$

Equations (18.7) and (18.8) give the probabilities for the initial and final states and these contain two terms corresponding to pure exponential decay and a third term which is a modulated decay. The results are in agreement with those obtained by the methods of Heitler-Ma [19] and Goldberger-Watson (Chapter 5).

B) Our second example is that of a two-level atom in an external field. We write the external perturbation in the form

$$H_{ext}(t) = -\tfrac{1}{2}|d|\{S^+ \mathscr{E}(t) + S^- \mathscr{E}^*(t)\},$$

and the master equation (18.4) for the present case is

$$\partial\varrho/\partial t = -i\omega[S^z, \varrho] + \tfrac{1}{2}i|d|[S^+ \mathscr{E}(t) + S^- \mathscr{E}^*(t), \varrho]$$
$$-\gamma(S^+ S^- \varrho - 2S^- \varrho S^+ + \varrho S^+ S^-). \tag{18.11}$$

The equations of motion are

$$(\partial/\partial t)\langle S^+\rangle = +i\omega\langle S^+\rangle + i|d|\mathscr{E}^*(t)\langle S^z(t)\rangle - \gamma\langle S^+\rangle,$$
$$(\partial/\partial t)\langle S^z\rangle = -2\gamma\langle S^+ S^-\rangle + \tfrac{1}{2}i|d|\mathscr{E}\langle S^+\rangle - \tfrac{1}{2}i|d|\mathscr{E}^*\langle S^-\rangle, \tag{18.12}$$

which are rather familiar (equations of motion for a damped, driven two-level atom) and have been studied extensively [97]. It has not been possible to solve (18.12) for the arbitrary time dependence of $\mathscr{E}(t)$. We present the solution for the case of a harmonically varying field

$$\mathscr{E}(t) = \mathscr{E}_0 e^{-i\omega_0 t}, \tag{18.13}$$

then on transforming to the rotating frame reduce (18.12) to

$$(\partial/\partial t)\langle S^+\rangle = i\Delta\langle S^+\rangle - \gamma\langle S^+\rangle + 2i\alpha\langle S^z\rangle,$$
$$(\partial/\partial t)\langle S^z\rangle = -2\gamma\langle S^+ S^-\rangle + i\alpha\langle S^+\rangle - i\alpha\langle S^-\rangle, \tag{18.14}$$
$$\alpha = \tfrac{1}{2}|d|\mathscr{E}_0, \qquad \Delta = \omega - \omega_0.$$

These equations are easily solved by taking the Laplace transforms:

$$\psi = \begin{pmatrix} \langle S^+\rangle \\ \langle S^-\rangle \\ \langle S^z\rangle \end{pmatrix}, \qquad I = \begin{pmatrix} 0 \\ 0 \\ 1 \end{pmatrix}, \qquad \hat\psi(z) = \mathscr{A}[\psi(0) - (\gamma/z)I]f^{-1}(z),$$

$$f(z) = 4\alpha^2(\gamma + z) + (z + 2\gamma)(z + \gamma - i\Delta)(z + \gamma + i\Delta),$$
$$A_{12} = A_{21} = 2\alpha^2,$$
$$A_{11} = 2\alpha^2 + (z + 2\gamma)(z + \gamma + i\Delta),$$
$$A_{22} = 2\alpha^2 + (z + 2\gamma)(z + \gamma - i\Delta), \tag{18.15}$$
$$A_{33} = \Delta^2 + (z + \gamma)^2,$$
$$A_{13} = 2A_{31} = -2\alpha\Delta + 2i\alpha(z + \gamma),$$
$$A_{23} = 2A_{32} = -2\alpha\Delta - 2i\alpha(z + \gamma).$$

The steady-state solution is given by

$$\langle S^+\rangle_{st} = -i\alpha(\gamma + i\Delta)\{2\alpha^2 + \gamma^2 + \Delta^2\}^{-1}$$
$$\langle S^z\rangle_{st} = -\tfrac{1}{2}\{1 + 2\alpha^2(\gamma^2 + \Delta^2)^{-1}\}^{-1}. \tag{18.16}$$

In the limit $\alpha \to 0$ these results reduce to those of Chapter 10. For the static field ($\omega_0 = 0$) the steady-state value of the atomic energy is

$$\omega \langle S^z \rangle_{st} = -\tfrac{1}{2}\omega \{1 + 2\alpha^2(\gamma^2 + \omega^2)^{-1}\}^{-1} \approx -\tfrac{1}{2}\omega(1 + 2\alpha^2/\omega^2) \quad \text{since } \gamma \ll \omega;$$

this is also equal to the ground-state energy of the atom in presence of a weak static field since the second-order energy correction to the ground state is $-(1/\omega)|\langle 2|H_{ext}|1\rangle|^2 = (\alpha^2/\omega)$ (the exact value of the ground-state energy is $-\tfrac{1}{2}\omega\{1 + 4\alpha^2/\omega^2\}^{\frac{1}{2}}$). If we had taken $\mathscr{E}_0$ to have a *random phase*, then $\langle S^+ \rangle_{st} = 0$.

For the case of a field at resonance $\varDelta = 0$, the roots of $f(z) = 0$ are

$$z = -\gamma, \quad -\tfrac{3}{2}\gamma \pm i\mu, \quad \mu = \tfrac{1}{2}(16\alpha^2 - \gamma^2)^{\frac{1}{2}}. \tag{18.17}$$

The zeros of $f(z)$ are simple so that the Laplace transform is easily inverted for calculating the time dependence of the dipole moment, energy, etc. For example, the mean value of the energy (subject to the initial condition $\varrho(0) = |1\rangle \langle 1|$) is given by

$$\langle S^z(t) \rangle = \langle S^z \rangle_{st} + \tfrac{1}{4}(\gamma^2 + 2\alpha^2)^{-1} e^{-\frac{3}{2}\gamma t}$$
$$\cdot \{[2(\gamma^2 + \alpha^2) - (\gamma/i\mu)(\gamma^2 + 5\alpha^2)] e^{i\mu t} + \mu \to -\mu\}, \tag{18.18}$$

which shows modulated experimental decay. The mean number of photons in any mode is given by (7.23) and in the present model there is no steady-state value for $N_{ks}(t)$, as it contains terms proportional to time t. However, the rate σ_{ks} of change of photons in any mode ks is finite. On using (18.15) it is easy to calculate all the relevant functions f, h which appear in (7.29) and a straightforward but tedious calculation leads to

$$\sigma_{ks}(\infty) = \frac{8|g_{ks}|^2\alpha^4\gamma(2\alpha^2 + 4\gamma^2 + \varDelta_0^2)}{(\gamma^2 + 2\alpha^2)(\gamma^2 + \varDelta_0^2)[9\gamma^2\varDelta_0^2 + (4\alpha^2 + 2\gamma^2 - \varDelta_0^2)^2]}, \quad \varDelta_0 = \omega - \omega_{ks}. \tag{18.19}$$

For fields that are such that $\alpha \gg \gamma$ (18.19) can be written as a sum of three Lorentzians centered at $\omega = \omega_{ks}$, $\omega_{ks} \pm 2\alpha$ and with half-widths γ, $\tfrac{3}{2}\gamma$, respectively, which is the usual dynamic Stark-effect triplet. The result so obtained for $\sigma_{ks}(\infty)$ agrees only qualitatively[23] with the results

---

[23] Mollow [97] has also calculated the power spectrum of the emitted light by taking it proportional to $\langle S^+(t) S^-(t') \rangle$, where $\langle \rangle$ refers to the *steady-state* averaging and thus the correlation function is a function only of $(t - t')$. The power spectrum which we calculate in § 10, 13, etc. is a function of both arguments. For the case when there is no external field and for a single two-level atom, one can easily show that the *steady-state* correlation $\langle S^+(t) S^-(t') \rangle$ vanishes and hence, according to Mollow, the power spectrum vanishes in contradiction to the results of § 10. In our opinion the correlation functions of Mollow are relevant in studying the *linear response* of the two-level atom (damped) to external perturbations but not the spontaneous emission problem.

existing in the literature [97 to 99]. It is obvious that in our model $\langle a_{ks}^+ a_{ks} \rangle$ does not possess a limiting value because the external field is always acting on the system and this excites the system to emit many photons.

We also consider briefly what happens if we use a Wigner-Weisskopf type of calculation. We make the following ansatz for the wave function

$$\psi(t) = b_0^+ |1, \{0\}\rangle \, e^{-iE_1 t} + b_0^- |2, \{0\}\rangle \, e^{-iE_2 t}$$
$$+ \sum_k b_k^+ |1, \{k\}\rangle \, e^{-iE_1 t - i\omega_k t} + \sum_k b_k^- |2, \{k\}\rangle \, e^{-iE_2 t - i\omega_k t}, \qquad (18.20)$$

where we have restricted ourselves to the consideration of one-photon transitions. The equations of motion for $b_0^\pm, b_k^\pm$ are

$$i\dot{b}_0^+ = \alpha b_0^- + \sum_k g_k b_k^- \, e^{i(\omega - \omega_k)t}, \quad i\dot{b}_0^- = \alpha b_0^+ ,$$
$$i\dot{b}_k^- = g_k^* b_0^+ \, e^{-i(\omega - \omega_k)t} + \alpha b_k^+ , \quad i\dot{b}_k^+ = \alpha b_k^- . \qquad (18.21)$$

We assume that the atom was initially in the excited state $b_0^+(0) = 1$, and all other amplitude coefficients vanish. On taking the Laplace transform, we obtain from (18.21)

$$\hat{b}_0^-(z) = -i\alpha z^{-1} \hat{b}_0^+(z) ,$$
$$\hat{b}_k^+(z) = -i\alpha z^{-1} \hat{b}_k^-(z) ,$$
$$\hat{b}_k^-(z) = -ig_k^* z(z^2 + \alpha^2)^{-1} \hat{b}_0^+(z + i\Delta_k) , \qquad (18.22)$$
$$\hat{b}_0^+(z) = z \left\{ (z^2 + \alpha^2) + z \sum_k |g_k|^2 (z - i\Delta_k) \left[\alpha^2 + (z - i\Delta_k)^2 \right]^{-1} \right\}^{-1} ,$$

$$\Delta_k \equiv \omega - \omega_k .$$

We now approximate $\hat{b}_0^+(z)$ by

$$\hat{b}_0^+(z) \approx z \{ z^2 + \alpha^2 + z(\gamma + i\Omega) \}^{-1} , \qquad (18.23)$$

where γ and Ω are the damping and the Lamb shift in the absence of the applied field. This approximation ignores the interference effects [cf. discussion following (18.3)]. With (18.23) one finds that

$$\hat{b}_k^-(z) \approx \frac{-ig_k^* z(z + i\Delta_k)}{(z^2 + \alpha^2) \left[(z + i\Delta_k)^2 + \alpha^2 + \gamma(z + i\Delta_k) \right]} , \qquad (18.24)$$

$$b_0^+(t) \approx (z_+ - z_-)^{-1} \{ z_+ \, e^{z_+ t} - z_- \, e^{z_- t} \} , \qquad (18.25)$$

$$z_\pm = -\tfrac{1}{2}(\gamma + i\Omega) \pm \tfrac{1}{2} \{ (\gamma + i\Omega)^2 - 4\alpha^2 \}^{\frac{1}{2}} . \qquad (18.26)$$

The probability that there is a photon in the mode ks is

$$p_{ks}(t) = |b_{ks}^+(t)|^2 + |b_{ks}^-(t)|^2 . \qquad (18.27)$$

Since $\hat{b}_{ks}(z)$ has poles at $z = \pm i\alpha$, $z_{\pm} - i\Delta_k$, it is clear that for $t \to \infty$ only the poles at $z = \pm i\alpha$ will contribute and hence

$$p_{ks}(\infty) \tag{18.28}$$

$$= \tfrac{1}{2}|g_{ks}|^2 |z_+ - z_-|^{-2} \left\{ \left| \frac{z_+}{(\Delta_k + \alpha + iz_+)} - \frac{z_-}{(\Delta_k + \alpha + iz_-)} \right|^2 + \alpha \to -\alpha \right\}.$$

The spectrum of the emitted radiation as given by (18.28) agrees with that obtained by Stroud [99, Eq. (35)] even though he treated the external field quantum mechanically whereas we treated it classically, although Stroud did assume that the excitation of the external field (single-mode field in the Fock state) was very high. It is presumably for this reason that we obtain the same result, since for large excitations the classical limit is expected to hold. In the limit of strong external fields $(\alpha \gg \gamma)$ (18.28) reduces to

$$p_{ks}(\infty) \approx \tfrac{1}{2}|g_{ks}|^2 \left\{ ([\gamma^2/4 + (\omega - \omega_{ks} + 2\alpha + \Omega/2)^2]^{-1} + \alpha \to -\alpha) \right.$$
$$\left. + 2(\gamma^2/4 + (\omega - \omega_{ks} + \Omega/2)^2)^{-1} \right\}. \tag{18.29}$$

The spectrum thus consists of three Lorentzians, as before, but now the widths and shifts are different − both the line width and the line shift have half the value of the widths and shifts in the absence of the applied field. Note that in obtaining (18.29) we made an important approximation, i.e. we truncated the problem by restricting it to one-photon transitions, and it is for this reason that there exists a steady state for $N_{ks}(\infty)$, which is also given by (18.29). In view of our result (18.19), it seems that the one-photon approximation is a poor approximation for very long times but is probably a good one for time intervals such that $1/\omega < t < 1/\alpha$.

We have so far treated the external fields as classical fields. We will now briefly discuss the case of a quantized external field (the black-body case has already been discussed in Chapter 17).

We assume that a coherent driving field is present so that the initial state of the field is given by

$$\varrho_R(0) = |\{z_{ks}\}\rangle \langle\{z_{ks}\}|, \tag{18.30}$$

where $|\{z_{ks}\}\rangle$ is a coherent state corresponding to the (ks) mode of the field. We are interested in deriving an equation of motion for the atomic-density operator. The projection operator $\mathscr{P}$ is given by (6.9)

$$\mathscr{P}\ldots = \varrho_R(0)\,\mathrm{Tr}_R\ldots,$$

with $\varrho_R(0)$ now given by (18.30). We will also restrict ourselves to the case of one two-level atom. The master equation is given by (6.25). We consider the master equation only in the Born approximation, i.e. we replace $U(\tau)$ in (6.25) by unity. For the state (18.30) the term $\mathscr{P}\mathscr{L}_{AR}\mathscr{P}\varrho_{A+R}$ also makes a contribution, i.e. (6.24) no longer holds. A straightforward calculation also shows that

$$\mathscr{P}\mathscr{L}_{AR}U_0(\tau)(1 - \mathscr{P})\mathscr{L}_{AR}\mathscr{P}\varrho_{A+R}(t - \tau)$$

does *not* depend on the *excitation amplitude*. One can make the Markov approximation, for the same reason as in the case of spontaneous emission, in the absence of any external field. One finds that the reduced-density operator obeys the equation

$$\partial \varrho / \partial t = -i[H_{ext}, \varrho] - i\omega[S^z, \varrho] - \gamma(S^+ S^- \varrho - 2S^- \varrho S^+ + \varrho S^+ S^-), \quad (18.31)$$

where

$$H_{ext} = -\boldsymbol{d} \cdot (S^+ + S^-) \, \mathscr{E}_0(t), \qquad (18.32)$$

$$\mathscr{E}_0(t) = i\Sigma (2\pi c k/L^3)^{\frac{1}{2}} z_{ks} \, \boldsymbol{\varepsilon}_{ks} \, e^{-i\omega_{ks}t} + \text{H.C.} \qquad (18.33)$$

Thus the effect of a quantized field in a coherent state is the same (in Born approximation) as the effect of a classical driving field [cf. (18.11)]. In particular, for the case of a field with only one mode in a coherent state, the equations of motion are identical to (18.14) if the RWA is also made.

It should be borne in mind that the master equation (6.25) is non-linear with respect to the initial state of the field, so that the case of a more general field as described by the density operator

$$\varrho_R(0) = \int \Phi(\{z_{ks}\}) |\{z_{ks}\}\rangle \langle\{z_{ks}\}| d^2(\{z_{ks}\}), \qquad (18.34)$$

cannot be studied simply by averaging (18.31) with respect to $\Phi(\{z_{ks}\})$. For a coherent field with random phases, the master equation is the same as in the presence of black-body radiation (§ 17); the parameter $\langle n_{ks} \rangle$ should be replaced by $|z_{ks}|^2$, whereupon the temperature-dependent Lamb shift becomes field-dependent. The spontaneous emission in the case when only one mode of the field is excited initially to a Fock state is much more involved; in such situations one has to use a non-Markovian master equation.

Finally, the master equation for a small sample driven by an external (classical) field is

$$\partial \varrho / \partial t = -i\omega[S^z, \varrho] - i\mathscr{E}[S^+, \varrho] - i\mathscr{E}^*[S^-, \varrho]$$
$$- \gamma(S^+ S^- \varrho - 2S^- \varrho S^+ + \varrho S^+ S^-), \qquad (18.35)$$

where we have ignored, as before, the effect of the dipole–dipole coupling term. This master equation can also be handled by the techniques of Chapter 14. The Langevin equations (14.1) are modified to

$$\dot{z}_i = -i\omega z_i - i\mathscr{E} + 2i\mathscr{E} |z_i|^2 - \gamma \sum_j z_j + 2\gamma |z_i|^2 \sum_{j \neq i} z_j. \qquad (18.36)$$

For the variables A and D [defined by (14.2)] we have the equations

$$\dot{D} = -i\omega D - iN\mathscr{E} + 2i\mathscr{E} A - \gamma ND + 2\gamma AD, \qquad (18.37)$$

$$\dot{A} = -2\gamma DD^* + i\mathscr{E}^* D - i\mathscr{E} D^*. \qquad (18.38)$$

For the case of a monochromatic field at resonance these Langevin equations can be solved by transforming them to the rotating frame. The solutions are rather involved and will be discussed elsewhere[24].

---

[24] The case of a three-level atom in an external field is discussed by Fain and Khanin [100].

Appendix A. Role of Rotating-Wave Approximation in Spontaneous Emission

The rotating-wave approximation (RWA) has conventionally been employed to treat problems in quantum optics involving the interaction of radiation and matter. RWA is equivalent to using the Hamiltonian (2.24) in place of (2.12). It is usually believed that the effect of the counter-rotating terms is negligible, provided the interaction between the radiation and matter is weak. For the case of a spin $-\frac{1}{2}$ system in a magnetic field, it was shown by Bloch and Siegert [101] that the effect of the counter-rotating terms is indeed negligible provided the field is not too strong. They also worked out the corrections to the transition probabilities. Similar results have been obtained for the case of a two-level atom in an external electric field by Autler and Towns [102] and by others [103, 104]. For a single-mode quantized electric field interacting with the two-level atom, results similar to those of Bloch and Siegert are again obtained [25]. In this appendix we discuss the role of RWA in spontaneous emission [32].

We first consider the case of a single two-level atom. The master Eq. (6.30) in the Markovian approximation for a single two-level atom reduces to

$$\partial \varrho/\partial t + i(\omega + \Omega_+)\,[S^z, \varrho] + i\Omega_+(S^+ \varrho S^+ - S^- \varrho S^-)$$
$$+ \gamma(S^+ S^- \varrho - 2S^- \varrho S^+ + \varrho S^+ S^- - S^+ \varrho S^+ - S^- \varrho S^-) = 0, \tag{A.1}$$

where

$$\gamma = \tfrac{2}{3}(\omega/c)^3 \,|d|^2 \,,$$

$$\Omega_+ \equiv \Omega_{ii} = \sum_{ks} |g_{ks}|^2 \,\{(\omega - \omega_{ks})^{-1} + (\omega + \omega_{ks})^{-1}\} \tag{A.2}$$

$$= \tfrac{2}{3}(|d|^2/\pi)\int k^3 \, dk\{(k + k_0)^{-1} - (k - k_0)^{-1}\}\,,$$

which, on taking self-interactions into account and on renormalization, becomes [cf. Eq. (6.48)]

$$\Omega_+ = -(\gamma/\pi)\ln\{|\omega_c/\omega - 1|\,(\omega_c/\omega + 1)\}\,. \tag{A.3}$$

In deriving (A.1) we used the commutation relations appropriate to a spin $-\frac{1}{2}$ system, i.e.

$$[S^+ S^-, \varrho] = -[S^- S^+, \varrho] = [S^z, \varrho]\,, \qquad S^+ S^+ = S^- S^- = 0\,. \tag{A.4}$$

If we transform (A.1) to the interaction picture, then we obtain

$$\partial \varrho_I/\partial t + i\Omega_+(S^+ \varrho_I S^+ \, e^{2i\omega_0 t} - \text{H.C.})$$
$$+ \gamma(S^+ S^- \varrho_I - S^- \varrho_I S^+ - S^+ \varrho_I S^+ \, e^{2i\omega_0 t} + \text{H.C.})\,, \tag{A.5}$$

where ω_0 is the renormalized frequency ($\omega_0 = \omega + \Omega_+$). On making RWA on the master Eq. (A.5), i.e. neglecting rapidly oscillating terms

---

[25] The present author used the method of time averaging [105] to obtain such results; similar results have also been obtained by Walls [106] who used the formalism of [103].

like $S^+ \varrho_I S^+ e^{2i\omega_0 t}$, we obtain

$$\partial\varrho/\partial t + i\omega_0[S^z, \varrho] + \gamma(S^+ S^- \varrho - 2S^- \varrho S^+ + \varrho S^+ S^-) = 0, \tag{A.6}$$

where we have also transformed back to the Schrödinger picture. The master equation (A.6) is the one studied in detail in Chapter 10. On the other hand, if we work with the Hamiltonian (2.24) obtained from (2.12) by making RWA, then a straightforward calculation shows that the master equation for the reduced-density operator is

$$\partial\varrho/\partial t + i(\omega + \Omega'_+)[S^z, \varrho] + \gamma(S^+ S^- \varrho - 2S^- \varrho S^+ + \varrho S^+ S^-) = 0, \tag{A.7}$$

where

$$\Omega'_+ = \tfrac{2}{3}(|d|^2/\pi)\int k^3 \, dk(k_0 - k)^{-1}. \tag{A.8}$$

The master equation (A.7) is identical to (A.6) except that the numerical value of Ω_+ is now different. It should be noted that Ω'_+ is simply the shift of the excited state of the atom. The shift of the ground state is missing from (A.7), mainly due to the virtual transitions which are automatically excluded by the Hamiltonian (2.24). The master equation (A.6) obtained by making RWA on the master equation rather than on the Hamiltonian does include the shift of the ground state. These remarks make it clear that RWA on the original Hamiltonian is not same as RWA on the master equation and that one should make RWA on the final equations of motion.

The counter-rotating terms such as $S^+ \varrho S^+$ in (A.1) are not important because $\gamma \ll \omega$. From (A.1) the equations of motion for the dipole moment are

$$(\partial/\partial t)\langle S^\pm\rangle = \pm i\omega_0\langle S^\pm\rangle - \gamma\langle S^\pm\rangle + \gamma(1 \pm i\Omega_+/\gamma)\langle S^\mp\rangle. \tag{A.9}$$

The general solution of (A.9) is

$$\langle S^-(t)\rangle = e^{-\gamma t} \cos\{(\omega_0^2 - \gamma^2 - \Omega_+^2)^{\frac{1}{2}} t\} \langle S^-(0)\rangle$$
$$+ (\gamma - i\Omega_+)(\omega_0^2 - \gamma^2 - \Omega_+^2)^{-\frac{1}{2}} e^{-\gamma t} \sin\{(\omega_0^2 - \gamma^2 - \Omega_+^2)^{\frac{1}{2}} t\} \langle S^+(0)\rangle. \tag{A.10}$$

Usually in an optical experiment one is not measuring $\langle S^\pm(t)\rangle$ but a time average of $\langle S^\pm(t)\rangle$ over several optical periods. It is easily seen that the time average of (A.10) over several optical cycles is same as (10.4) if $\gamma \ll \omega$; this also happens to be one of the limits of the validity of the master equation. We also discuss briefly the correction terms to (10.4). To obtain such correction terms, one can use several methods. Here we use the Bogoliubov-Mitropolsky method of time averaging [105] which is well known in mechanics. We can apply it either on the master equation (A.5) or on the Eq. (A.9). The dipole moment in the interaction picture satisfies

$$(\partial/\partial t)\langle S^\pm\rangle = -\gamma\langle S^\pm\rangle + \gamma(1 \pm i\Omega_+/\gamma) e^{\mp 2i\omega_0 t}\langle S^\mp\rangle, \tag{A.11}$$

which we rewrite in the form

$$\dot{\psi} = \gamma(H_1 + H_2(t)) \, \psi \, , \qquad \psi \equiv \begin{pmatrix} \langle S^+ \rangle \\ \langle S^- \rangle \end{pmatrix},$$

$$H_1 = -1 \, , \quad H_2(t) = \begin{pmatrix} 0 & (1 + i\Omega_+/\gamma) \, e^{-2i\omega_0 t} \\ (1 - i\Omega_+/\gamma) \, e^{+2i\omega_0 t} & 0 \end{pmatrix}. \tag{A.12}$$

The time average of $H_2(t)$ (denoted by a bar) is zero. The method of time averaging consists in writing

$$\psi = \phi + \gamma F_1 + \gamma^2 F_2 + \cdots \, , \qquad \partial\phi/\partial t = \gamma P_1 + \gamma^2 P_2 + \cdots \, , \tag{A.13}$$

where ϕ is the time-averaged part of ψ (experimentally detected part of ψ). On substituting (A.13) in (A.12) and equating the terms of each order in γ, and on calculating P_1, P_2 etc. by the requirement $\overline{P}_1 = P_1$, $\overline{\phi} = \phi$, we find that the dipole moment in the lowest order in (γ/ω_0) satisfies the equation

$$\langle \dot{S}^{\pm} \rangle = -\gamma[1 \pm (i\gamma/2\omega_0)(1 + \Omega_+^2/\gamma^2)] \langle S^{\pm} \rangle \, , \tag{A.14}$$

whose solution is

$$\langle S^{\pm}(t) \rangle = \exp\{\pm i[\omega_0 - (\gamma^2/2\omega_0)(1 + \Omega_+^2/\gamma^2)] \, t - \gamma t\} \langle S^{\pm}(0) \rangle \, . \tag{A.15}$$

On comparing (A.15) with (10.4) we see that the effect of the counter-rotating terms in (A.11) is to give rise to additional frequency-shift terms and to this order the damping term γ is unaffected. The frequency-shift terms of the form γ^2/ω_0, which are the analog of the Bloch-Siegert shifts well known in magnetic resonance theory, have also been discussed by Arnous and Heitler [107] using a different formalism. Such frequency-shift terms are, however, negligibly small in the present context.

We now discuss the case of many two-level atoms. It should be noted that the master equation obtained by using (2.24) is given by

$$\partial\varrho/\partial t + i \sum_i (\omega + \Omega'_+) \, [S_i^z, \varrho] + i \sum_{i \neq j} \Omega'_{ij} [S_i^+ S_j^- , \varrho]$$
$$+ \sum_{ij} \gamma_{ij} (S_i^+ S_j^- \varrho - 2 S_j^- \varrho S_i^+ + \varrho S_i^+ S_j^-) = 0 \, , \tag{A.16}$$

where γ_{ij} is given by (6.45) and Ω'_+ by (A.8). The cooperative frequency shift term Ω'_{ij} is now

$$\Omega'_{ij} = -(|d|^2/4\pi^2) \int k^3 \, dk(k - k_0)^{-1} \int d\Omega' \, e^{i\mathbf{k} \cdot \mathbf{r}_{ij}} \sin^2 \theta' \, . \tag{A.17}$$

The master equation (A.16) has the same structure as (6.40) but the numerical values of the parameters corresponding to the shift terms are different. Ω'_{ij} as given by (A.17) is equal to $\frac{1}{2}\Omega_{ij} + \frac{1}{2} \, \mathrm{Im} \, Q_E(r_{ij}, \omega)$, with Q_E defined by (6.71 b) and does not diverge contrary to the result of [108]. RWA on the master equation leads to the right shift. For a pair of atoms, say one and two, the term $S_1^+ S_2^-$ has non-vanishing matrix elements only between the states $|1, 2\rangle$ and $|2, 1\rangle$ ($|1\rangle$ and $|2\rangle$ representing the excited and the ground states, respectively). The term Ω_{12}' takes into account only the resonant transition

$$|1, 2, \{0\}\rangle \rightarrow |2, 2, \{k\}\rangle \rightarrow |2, 1, \{0\}\rangle,$$

with $\{0\}$ representing the vacuum state of the field, whereas Ω_{12} also has a contribution from the virtual transition

$$|1, 2, \{0\}\rangle \rightarrow |1, 1, \{k\}\rangle \rightarrow |2, 1, \{0\}\rangle.$$

Either second-order perturbation theory or the master equation (6.40) shows (cf. Chapter 9) that the cooperative frequency shift for the states $|1,2\rangle$, $|2,1\rangle$ vanishes [26]. The cooperative frequency shift is nonvanishing only for the linear superposition of these states, viz $(1/\sqrt{2})(|1, 2\rangle \pm |2, 1\rangle)$. We also remark that the cooperative frequency shift is not *additive*. It is clear from the above discussion that, when ever questions of frequency shifts involving Lamb shifts, cooperative shifts, Casimir-Polder forces [109], or higher-order forces are considered, the RWA should not be made on the Hamiltonian. It is best to make RWA on the master equation as discussed above, although we have not considered higher-order effects (beyond dipole approximation and second-order perturbation theory) in the present article.

Appendix B.
Mori's Generalized Langevin-Equation Treatment of Spontaneous Emission

Mori has shown in a general way how Heisenberg equations of motion can be written in the form of Langevin equations. If G_i is the set of system operators, then Mori [65] introduces the projection operator

$$\mathscr{P} \ldots = \sum_i (\ldots, G_i)\, G_i(0)/(G_i, G_i), \tag{B.1}$$

where the scalar product is to be chosen according to the problem. A formal solution of the Heisenberg equations of motion is

$$G_i(t) = \exp\{i\mathscr{L}t\}\, G_i(0) = \exp\{i\mathscr{P}\mathscr{L}t + i(1 - \mathscr{P})\,\mathscr{L}t\}\, G_i(0),$$

which on using the identity

$$\exp\{i(A + B)\, t\} = \exp(iAt) + \int_0^t d\tau\, e^{i(A+B)\tau} iB\, e^{iA(t-\tau)}, \tag{B.2}$$

---

[26] By reading [108] it is not clear to the present author what initial state was used by Knight and Allen, although they mention that atom one is excited and atom two unexcited. We suppose that they take a symmetrized or unsymmetrized state, for otherwise the cooperative shift vanishes.

becomes

$$G_i(t) = \exp\{i(1 - \mathscr{P})\mathscr{L}t\}\, G_i(0) + \int_0^t d\tau\, e^{i\mathscr{L}\tau}\, i\mathscr{P}\mathscr{L}$$
$$\cdot \exp\{i(1 - \mathscr{P})\mathscr{L}(t - \tau)\}\, G_i(0)\,. \tag{B.3}$$

On taking the time derivative of (B.3), and on simplification, we obtain

$$\dot{G}_i + \int_0^t d\tau \sum_j \Phi_{ij}(t - \tau)\, G_j(\tau) = F_i(t) + \sum_j i\omega_{ij} G_j(t)\,, \tag{B.4}$$

where

$$F_i(t) = \exp\{i(1 - \mathscr{P})\mathscr{L}t\}\,(1 - \mathscr{P})\, i\mathscr{L}\, G_i(0)\,, \tag{B.5a}$$

$$\Phi_{ij}(\tau) = (\mathscr{L}\, e^{i(1 - \mathscr{P})\mathscr{L}\tau}(1 - \mathscr{P})\,\mathscr{L}\, G_i, G_j)/(G_j, G_j)\,, \tag{B.5b}$$

$$\omega_{ij} = -i(\dot{G}_i(0), G_j(0))/(G_j, G_j)\,. \tag{B.5c}$$

(B.4) is the desired Langevin equation, which is linear. The second term on the right-hand side of (B.4) gives the unperturbed motion. $F_i(t)$ are the random forces and $\Phi_{ij}(\tau)$ are the memory kernels. All the complications of the interaction are contained in Φ_{ij} and F_i. In the case that Mori considered the initial density operator was taken to be an equilibrium one, i.e. $\varrho = f(H)$; then if we define the scalar product (which is to be chosen so that $(G, G) \neq 0$)

$$(A, B) = \mathrm{Tr}\{\varrho A^+ B\}\,, \tag{B.6}$$

it can be shown that Φ and F are related by the so-called second fluctuation-dissipation theorem

$$\Phi_{ij}(\tau) = (F_j(0), F_i(\tau))\,(G_j, G_j)^{-1}\,, \tag{B.7}$$

and this is a major simplification. The memory kernel Φ is closely related to Dyson's self-energy operator. For the calculation of shifts and widths we can very well restrict ourselves to equilibrium situations. The memory kernel can be evaluated by continued fractions.

For the nonequilibrium situation, as in the case of spontaneous emission, a simple relation like (B.7) does not appear to hold. There is another problem with Langevin equations of the form (B.4); this arises when one tries to approximate the memory kernel and the random force. If we write (B.4) for the operator S_i^-, we see that the structure of (B.4) is very different from that of (8.31) with Q standing for S_i^-, although both equations are exact and are equivalent to each other. It appears to us that one can develop the operator $e^{i(1 - \mathscr{P})\mathscr{L}\tau}$ in powers of the interaction and truncate the series *only* if we choose for the operators G_i a *complete* set of atomic operators (in case a complete set is not chosen, one should handle such equations with great care!). In what follows we choose for the operators a complete set of atomic operators. We introduce the projection operator

$$\mathscr{P}\ldots = \text{Tr}_R \varrho_R(0)\ldots. \tag{B.8}$$

The Langevin equation and the random force are given by (B.4) and (B.5a), respectively, with

$$\omega_{ij} = -i\,\text{Tr}_R\varrho_R(0)\,\text{Tr}_A\{G_j^+\dot{G}_i\}\,, \tag{B.9a}$$

$$\Phi_{ij} = \text{Tr}_R\varrho_R(0)\,\text{Tr}_A\{G_j^+\,\mathscr{L}\,e^{i(1-\mathscr{P})\mathscr{L}\tau}(1-\mathscr{P})\,\mathscr{L}G_i\}\,. \tag{B.9b}$$

The random force F_i has the property

$$\mathscr{P}F_i(t) = 0 \Rightarrow \langle F_i(t)\rangle = 0\,, \tag{B.9c}$$

and therefore for the mean values we have

$$\langle\dot{G}_i\rangle + \int_0^t d\tau \sum_j \Phi_{ij}(t-\tau)\langle G_j(\tau)\rangle = i\sum_j \omega_{ij}\langle G_j(t)\rangle\,. \tag{B.10}$$

If (B.10) is approximated by

$$\langle\dot{G}_i\rangle + \sum_j\langle G_j(t)\rangle\int_0^\infty d\tau\,\Phi_{ij}^0(\tau) = i\sum_j \omega_{ij}\langle G_j(t)\rangle\,, \tag{B.11a}$$

where

$$\Phi_{ij}^0(\tau) = \text{Tr}_R\varrho_R(0)\,\text{Tr}_A\{G_j^+\,\mathscr{L}\,e^{i\mathscr{L}_0\tau}(1-\mathscr{P})\,\mathscr{L}G_i\}\,, \tag{B.11b}$$

and if one further makes the rotating-wave approximation, then (B.11a) turns out to be equivalent to the master equation. As before [cf. (8.37)], the evaluation of the correlation functions of the random force is very difficult, although it can be done, and it leads to the same result as the master-equation approach.

The harmonic-oscillator model is an exception to the above: it is possible to obtain exact Langevin equations. For a complete set of operators G_i we take

$$G_i = |m\rangle_i\,{}_i\langle n| = A_{mn}^{(i)}\,, \tag{B.12}$$

where $|m\rangle_i$ is the Fock state of the ith oscillator. Using (2.12) we find

$$\begin{aligned}
i(1-\mathscr{P})\,\mathscr{L}\,a_i &= i(1-\mathscr{P})\Big[\sum_j \omega a_j^+ a_j + \sum_{ks}\omega_{ks}a_{ks}^+ a_{ks}\\
&\quad + \Big\{\sum_{jks} g_{jks}a_{ks}(a_j^+ + a_j) + \text{H.C.}\Big\}, a_i\Big]\\
&= -i\sum_{ks} g_{iks}a_{ks} - i\sum_{ks} g_{iks}^* a_{ks}^+\,,
\end{aligned}$$

$$i(1-\mathscr{P})\,\mathscr{L}\,a_i^+ = i\sum_{ks} g_{iks}^* a_{ks}^+ + i\sum_{ks} g_{iks}a_{ks}\,. \tag{B.13}$$

We now write

$$e^{i(1-\mathscr{P})\mathscr{L}t} = e^{i(1-\mathscr{P})\mathscr{L}_0 t} + \int_0^t e^{i(1-\mathscr{P})\mathscr{L}\tau} i(1-\mathscr{P})\mathscr{L}_1 e^{i(1-\mathscr{P})\mathscr{L}_0(t-\tau)}\,d\tau, \quad (B.14a)$$

$$e^{i(1-\mathscr{P})\mathscr{L}_0 t} = e^{i\mathscr{L}_0 t} + \int_0^t e^{i(1-\mathscr{P})\mathscr{L}_0\tau}(-i\mathscr{P}\mathscr{L}_0)e^{i\mathscr{L}_0(t-\tau)}\,d\tau. \quad (B.14b)$$

Hence

$$e^{i(1-\mathscr{P})\mathscr{L}_0 t}a_{ks} = a_{ks}\,e^{-i\omega_{ks}t} - i\int_0^t e^{i(1-\mathscr{P})\mathscr{L}_0 t}\,\mathscr{P}[H_0, a_{ks}]\,e^{-i\omega_{ks}(t-\tau)}\,d\tau$$

$$= a_{ks}\,e^{-i\omega_{ks}t}, \quad (B.15)$$

the second term vanishes because $\mathscr{P}a_{ks} = 0$. From (B.15) and (B.14a) we obtain

$$e^{i(1-\mathscr{P})\mathscr{L}t}a_{ks} = a_{ks}\,e^{-i\omega_{ks}t} + \int_0^t e^{i(1-\mathscr{P})\mathscr{L}(t-\tau)}\,i(1-\mathscr{P})[H_1, a_{ks}]$$

$$\cdot e^{-i\omega_{ks}\tau}\,d\tau = a_{ks}\,e^{-i\omega_{ks}t}, \quad (B.16)$$

the second term vanishes because $(1-\mathscr{P})a_j = (1-\mathscr{P})a_j^+ = 0$. The Langevin equation for the operator a_i would be

$$\dot{a}_i = \sum_{mn}(a_i)_{mn}\dot{A}_{mn}^{(i)} = F_i(t) - i\omega a_i$$

$$- \int_0^t d\tau \sum_{pqj} \Phi_{pqj}^{(i)}(t-\tau) A_{pq}^{(j)}(\tau), \quad (B.17)$$

where on combining (B.16) and (B.5a) the random force F_i is given by

$$F_i(t) = -i\sum_{ks} g_{iks}a_{ks}\,e^{-i\omega_{ks}t} - i\sum_{ks} g_{iks}a_{ks}^+\,e^{i\omega_{ks}t}, \quad (B.18a)$$

and the memory kernel is

$$\Phi_{pqj}^{(i)} = -\mathrm{Tr}_R\varrho_R(0)\,\mathrm{Tr}_A\left\{A_{pq}^{(j)+}\left[H, \sum_{ks} g_{iks}a_{ks}\,e^{-i\omega_{ks}\tau} + \text{H.C.}\right]\right\}$$

$$= \sum_{ks} g_{iks}g_{jks}^*\,e^{-i\omega_{ks}\tau}\,_j\langle p|\,a_j^+ + a_j\,|q\rangle_j \quad (B.18b)$$

$$- \sum_{ks} g_{iks}^*g_{jks}\,e^{i\omega_{ks}\tau}\,_j\langle p|\,a_j^+ + a_j\,|q\rangle_j.$$

Hence (B.17) becomes

$$\dot{a}_i + i\omega a_i + \int_0^t d\tau \sum_{ksj} g_{jks}^*g_{iks}\,e^{-i\omega_{ks}\tau}[a_j(t-\tau) + a_j^+(t-\tau)]$$

$$- \int_0^t d\tau \sum_{ksj} g_{iks}^*g_{jks}\,e^{i\omega_{ks}\tau}[a_j(t-\tau) + a_j^+(t-\tau)] = F_i(t). \quad (B.19)$$

These are the exact Langevin equations, which we have derived by using the Mori formalism in order to show how it works. Needless to say, (B.19) can also be obtained much more simply by using the procedure of Chapter 8. The random force has the properties

$$\langle F_i(t) \rangle = 0, \langle F_i(t) F_j^+(t') \rangle = \sum_{ks} g_{iks} g_{jks}^* e^{-i\omega_{ks}(t-t')},$$

$$\langle F_j^+(t') F_i(t) \rangle = \sum_{ks} g_{jks} g_{iks}^* e^{i\omega_{ks}(t-t')}. \tag{B.20}$$

Finally we consider briefly the memory kernel $\Phi(\tau)$ for a single-atom problem. We take for G_i the operator S^- and introduce the projection operator

$$\mathcal{P}G = \mathrm{Tr}\{\varrho(0)(G^+ S^- + S^- G^+)\}/\mathrm{Tr}\{\varrho(0)(S^+ S^- + S^- S^+)\} S^-$$
$$= \mathrm{Tr}\{\varrho(0)(G^+ S^- + S^- G^+)\} S^-. \tag{B.21}$$

Then (B.4) shows that the mean (exact) equation of motion for the dipole moment is

$$\langle \dot{S}^- \rangle + i\omega \langle S^- \rangle + \int_0^t d\tau\, \Phi(\tau) \langle S^-(t-\tau) \rangle = 0, \tag{B.22}$$

where

$$\Phi(\tau) = \mathrm{Tr}\{\varrho(0)\,[(\mathscr{L}\,e^{i(1-\mathscr{P})\mathscr{L}\tau}(1-\mathscr{P})\,\mathscr{L}S^-)^+ S^-$$
$$+ S^-(\mathscr{L}\,e^{i(1-\mathscr{P})\mathscr{L}\tau}(1-\mathscr{P})\,\mathscr{L}S^-)^+]\}. \tag{B.23}$$

It is clear from (B.22) that the real and the imaginary parts of Φ give the decay constant and Lamb shift, respectively. So far (B.23) is also exact and can be written as

$$\Phi(\tau) = \langle \{\mathscr{A}^+, S^-\} \rangle_0, \tag{B.24}$$

where the average is with respect to the initial density operator and

$$\mathscr{A} = \mathscr{L}\,e^{i(1-\mathscr{P})\mathscr{L}\tau}(1-\mathscr{P})\,\mathscr{L}S^-. \tag{B.25}$$

The expression for the memory kernel (B.24) bears some resemblance to recent work of Bullough and Caudery [48] where the linear response theory has been considered. It reduces to their expression if the theory is worked out to the lowest order in the coupling constant, in which case $\mathscr{A}$ reduces to

$$\mathscr{A} \approx \mathscr{L}\,e^{i\mathscr{L}_0\tau}(1-\mathscr{P})\,\mathscr{L}S^-. \tag{B.26}$$

On substituting (B.26) in (B.24) we find after a trivial calculation

$$\Phi(\tau) = \langle \{d \cdot E(\tau), d \cdot E(0)\}_+ \rangle, \tag{B.27}$$

i.e. the memory function for the *dipole moment* is given by the mean value of the anticommutator of the free field operators at two different time points. Equations (B.22) and (B.24) are equivalent to the results of [48] where a fermion decorrelation technique has been used [see also our discussion following (6.70)]. It should be noted that (B.27) is also valid in the case when external fields are also present, provided they have a random phase so that terms like $\langle a_{ks} \rangle$, $\langle a_{ks}^+ \rangle$ are zero.

Appendix C. Steady-State Solution of the Master Equations from the Viewpoint of Microreversibility

In this appendix we discuss the steady-state solution of the master equation from the viewpoint of microreversibility. In the previous sections we have discussed the steady-state characteristics by solving the time-dependent master equation. The steady-state solution can be directly obtained by invoking microreversibility, if it holds. We have discussed this principle for open quantum Markovian systems elsewhere [110] and here we merely quote the result. It was found that a Markovian system, characterized by the master equation

$$\partial \varrho / \partial t = \mathscr{L} \varrho \,, \tag{C.1}$$

must have a steady-state solution ϱ_{st} such that

$$\varrho_{st} = \tilde{\varrho}_{st} \,, \tag{C.2}$$

$$\varrho_{st} \bar{\mathscr{L}} = \tilde{\mathscr{L}} \tilde{\varrho}_{st} \,, \tag{C.3}$$

in order for microreversibility to hold. In (C.2) the operator $\tilde{G}$ denotes the operator obtained by time reversal and $\bar{\mathscr{L}}$ and $\tilde{\mathscr{L}}$ are defined by

$$\mathrm{Tr}\{A(\mathscr{L}B)\} \equiv \mathrm{Tr}\{B(\bar{\mathscr{L}}A)\} \,, \tag{C.4}$$

$$\widetilde{\mathscr{L}A} \equiv \tilde{\mathscr{L}}\tilde{A} \,. \tag{C.5}$$

We first consider the master Eq. (6.53): the operators $\mathscr{L}$, $\bar{\mathscr{L}}$ and $\tilde{\mathscr{L}}$ are given by

$$\begin{aligned}
\mathscr{L}G = &-i\omega \sum_i [S_i^z, G] - i \sum_{i \neq j} \Omega_{ij}[S_i^+ S_j^-, G] \\
&- \sum_{ij} \gamma_{ij}(S_i^+ S_j^- G - 2S_j^- GS_i^+ + GS_i^+ S_j^-) \,,
\end{aligned} \tag{C.6}$$

$$\begin{aligned}
\bar{\mathscr{L}}G = &i\omega \sum_i [S_i^z, G] + i \sum_{i \neq j} \Omega_{ij}[S_i^+ S_j^-, G] \\
&- \sum_{ij} \gamma_{ij}(S_i^+ S_j^- G - 2S_i^+ GS_j^- + GS_i^+ S_j^-) \,,
\end{aligned} \tag{C.7}$$

$$\begin{aligned}
\tilde{\mathscr{L}}G = &i\omega \sum_i [S_i^z, G] + i \sum_{i \neq j} \Omega_{ij}[S_i^+ S_j^-, G] \\
&- \sum_{ij} \gamma_{ij}(S_i^+ S_j^- G - 2S_j^- GS_i^+ + GS_i^+ S_j^-) \,.
\end{aligned} \tag{C.8}$$

In obtaining (C.8) we used the relations

$$\tilde{S}_i^{\pm} = S_i^{\mp}, \quad \tilde{S}_i^z = S_i^z.$$

The detailed balance condition (C.3), which should hold as an operator identity, now leads to

$$i\omega \sum_i [S_i^z, \varrho_{st}] + i \sum_{i \neq j} \Omega_{ij} [S_i^+ S_j^-, \varrho_{st}] - \sum_{ij} \gamma_{ij} [S_i^+ S_j^-, \varrho_{st}] = 0, \qquad (C.9)$$

$$\sum_j \gamma_{ij} S_j^- \varrho_{st} = \sum_j \gamma_{ij} \varrho_{st} S_i^+ = 0. \qquad (C.10)$$

We have already remarked that the matrix γ_{ij} should be a semipositive-definite matrix in order that the semipositive definiteness of ϱ is preserved in time. We *assume* that the matrix γ_{ij} is not only semipositive-definite but *strictly positive-definite*. We can then invert the relations (C.10) to obtain

$$S_j^- \varrho_{st} = \varrho_{st} S_j^+ = 0 \quad \forall j. \qquad (C.11)$$

Once (C.11) is satisfied, the relation (C.9) is automatically fullfilled. Equation (C.11) implies that the steady state solution is

$$\varrho_{st} = \prod_i |2\rangle_i {}_i\langle 2|, \qquad (C.12)$$

i.e. each atom is found in its ground state in the steady state. We could not obtain the steady-state solution of (6.53) in the previous sections because the time-dependent solution is not known. However, we obtained the stationary-state solution here rather simply on grounds of micro-reversibility.

For the master Eq. (13.5) describing emission from a small sample, the matrix $\gamma_{ij}(=\gamma)$ is a semipositive-definite matrix and the condition (C.10) becomes

$$S^- \varrho_{st} = \varrho_{st} S^+ = 0, \qquad (C.13)$$

implying a steady-state solution of the form

$$\varrho_{st} = \sum \alpha_{SS'} |S, -S\rangle \langle S', -S'|. \qquad (C.14)$$

The condition (C.9) for a small sample reduces to

$$i\omega [\varrho_{st}, S^z] - \gamma [\varrho_{st}, S^+ S^-] - i\alpha_+ [\varrho_{st}, S^+ S^-] - i\alpha_- [\varrho_{st}, S^- S^+] = 0,$$

which on using (C.13) becomes

$$[\varrho_{st}, S^z] = 0. \qquad (C.15)$$

On combining (C.14) and (C.15) we find that the steady-state solution should have the form

$$\varrho_{st} = \sum \alpha_S |S, -S\rangle \langle S, -S|. \qquad (C.16)$$

In this case the detailed balance does not uniquely determine the stationary solution. Note, however, that for small samples S^2 is a constant of motion and therefore the steady-state solution is uniquely determined.

For the harmonic oscillator model, the stationary state solution is

$$\varrho_{st} = \prod_i |0\rangle_i \,_i\langle 0|, \tag{C.17}$$

if the matrix γ_{ij} is *positive-definite*; thus each oscillator is left in its ground state. For a small sample the analog of (C.13) is

$$D\varrho_{st} = \varrho_{st} D^+ = 0, \tag{C.18}$$

and thus the system is found in the ground state of the collective operator D.

We next consider the master equation for the degenerate three-level atom and restrict ourselves to the simplified case (15.31). A straightforward use of the conditions (C.2) and (C.3) shows that

$$\varrho_{st\,ij} = \varrho_{st\,ji}, \tag{C.19}$$

$$(A_{31} + A_{32})\varrho_{st} = \varrho_{st}(A_{13} + A_{23}) = 0, \quad [A_{11} + A_{22}, \varrho_{st}] = 0. \tag{C.20}$$

From (C.20) it follows that

$$\varrho_{11} + \varrho_{21} = \varrho_{12} + \varrho_{22} = \varrho_{13} = \varrho_{23} = \varrho_{31} = \varrho_{32} = 0. \tag{C.21}$$

Thus the microreversibility does not determine the steady-state solution uniquely in the sense that it leaves ϱ_{11} and ϱ_{22} arbitrary. The situation is similar to the case (C.16). Again in the present case there is a constant of motion (15.32) which enables us to obtain the steady-state solution uniquely

$$\varrho_{11}(\infty) = \varrho_{22}(\infty) = \alpha/4, \varrho_{12}(\infty) = \varrho_{21}(\infty) = -\alpha/4, \tag{C.22}$$

all other matrix elements of $\varrho(\infty)$ being zero. The parameter α is determined from the initial condition, for the symmetric excitation $\alpha = 0$, and the atom is left in the ground state.

The steady-state solution of the master equation in presence of external fields cannot be obtained because principle of microreversibility does not hold in general; one should therefore resort to other methods [111].

Acknowledgement. This work was done while the author was at Institute für Theoretische Physik, Universität Stuttgart. The author wishes to thank H. Haken and other members of the Institute for their hospitality. Some of the work reported here had been started while the author was at Rochester. The author would like to thank the US Air Force Office of Scientific Research for partial support during the preparation of this article and R.K.Bullough, J.H.Eberly, F.Haake, H.Haken, L.Mandel, P.Schwendimann, N.Rehler, E.Wolf and many others for very many stimulating conversations.

References

1. Weisskopf, V., Wigner, E.: Z. Physik **63**, 54 (1930); **65**, 18 (1931).
2. Weisskopf, V.: Ann. Physik **9**, 23 (1931); – Z. Physik **85**, 451 (1933).
3. Heitler, W., Ma, S. T.: Proc. Roy. Irish Acad. **52**, 109 (1949).
4. Heitler, W.: Quantum theory of radiation, 3. Ed., § 16. London: Oxford University Press.
5. Goldberger, M. L., Watson, K. M.: Collision theory. New York: John Wiley 1964; Chap. 8 and the references quoted in this chapter.
6. Low, F. E.: Phys. Rev. **88**, 53 (1952).
7. Chang, C. S., Stehle, P.: Phys. Rev. A **4**, 630 (1971).
8. Dicke, R. H.: Phys. Rev. **93**, 99 (1954).
9. Jaynes, E. T., Cummings, F. W.: Proc. IEEE **51**, 89 (1963).
10. Stroud, C. R., Jaynes, E. T.: Phys. Rev. A **1**, 106 (1970).
11. Crisp, M. D., Jaynes, E. T.: Phys. Rev. **179**, 1253 (1969).
12. Jaynes, E. T.: In: Mandel, L., Wolf, E. (Eds.): Proc. Third Rochester Conference on Coherence and Quantum Optics, p. 35. Plenum Publishing Corporation 1973.
13. Wessner, J. M., Anderson, D. K., Robiscoe, R. T.: Phys. Rev. Letters **29**, 1126 (1972). — Clauser, J. F.: Phys. Rev. A **6**, 49 (1972). — Hibbs, H. M.: In: Mandel, L., Wolf, E. (Eds.): Proc. Third Rochester Conference on Coherence and Quantum Optics, p. 83. Plenum Publishing Co. 1973.
14. Haken, H., Hübner, R., Zeile, K.: In: Kay, S. M., Maitland, A., (Eds.): Quantum optics, New York: Academic Press, 1970, p. 483.
15. Power, E. A., Zienau, S.: Phil. Trans. Roy. Soc. **251**, 427 (1959).
16. Feynman, R. P., Vernon, F. L., Hellwarth, R. W.: J. Appl. Phys. **28**, 49 (1957).
17. Lamb, W. E.: Phys. Rev. **85**, 259 (1952).
18. Czarnik, J. W., Fontana, P. R.: J. Chem. Phys. **50**, 4071 (1969).
19. Fontana, P. R., Lynch, D. J.: Phys. Rev. A **2**, 347 (1970). — Hearn, D. D., Fontana, P. R.: J. Chem. Phys. **51**, 1871 (1969).
20. Kroll, N. M.: In: Cohen-Tannoudji et al. (Eds.): Quantum optics and quantum electronics, p. 47. New York: Gordon and Breach 1965.
21. Goldhaber, A. S., Watson, K. M.: Phys. Rev. **160**, 1151 (1967).
22. Mower, L.: Phys. Rev. **142**, 799 (1966); **165**, 145 (1968).
23. Arecchi, F. T., Banfi, G. P., Fossati Bellani, V.: Nuovo Cimento **11** B, 276 (1972).
24. Lambropoulus, P.: Phys. Rev. **164**, 84 (1967).
25. Zwanzig, R. W.: In Brittin, W. E. (Ed.): Lectures in theoretical physics, Vol. III. New York: John Wiley 1961; Physica **33**, 119 (1964).
26. Argyres, P. N.: In: Brittin, W. E. (Ed.): Lectures in theoretical physics, Vol. VIIIA, p. 183. Boulder: University of Colorado Press 1966.
27. Agarwal, G. S.: In: Wolf, E. (Ed.): Progress in optics, Vol. XI, p. 1. Amsterdam: North Holland Publishing Co. 1973.
28. Haake, F.: Springer Tracts Mod. Phys. **66**, 98 (1973).
29. Agarwal, G. S.: Phys. Rev. A **2**, 2038 (1970).
30. Agarwal, G. S.: Phys. Rev. A **3**, 1783 (1971).
31. Agarwal, G. S.: Phys. Rev. A **4**, 1791 (1971).
32. Agarwal, G. S.: Phys. Rev. A **4**, 1778 (1971); A **7**, 1195 (1973).
33. Agarwal, G. S.: In: Mandel, L., Wolf, E. (Eds.): Proc. Third Rochester Conference on Coherence and Quantum Optics, p. 157. New York: Plenum Publishing Corporation 1973.
34. Lehmberg, R. H.: Phys. Rev. A **2**, 883 (1970).
35. Belavin, A. A., Zeldovich, B. Ya., Perelomov, A. M., Popov, V. S.: Sov. Phys. JETP **56**, 264 (1969).

36. Schwinger, J.: In: Schwinger et al. (Eds.): Quantum theory of angular momentum, p. 229. New York: Academic Press 1965.
37. Agarwal, G. S., Wolf, E.: Phys. Rev. D 2, 2161, 2187 (1970); and references therein.
38. Hamilton, J.: Proc. Phys. Soc. 62, 12 (1948). — Arecchi, F. T., Courtens, E.: Phys. Rev. A 2, 1730 (1970).
39. Davidson, R., Kozak, J. J.: J. Math. Phys. 11, 189 (1970).
40. Picard, R. H., Willis, C. R.: In: Mandel, L., Wolf, E. (Eds.): Proc. Third Rochester Conference on Coherence and Quantum Optics, p. 675. New York: Plenum Publishing Co. 1973; Phys. Rev. A 8, 1536 (1973).
41. Ackerhalt, J. R., Knight, P. L., Eberly, J. H.: Phys. Rev. Letters 30, 456 (1973).
42. Power, E. A.: Introductory quantum electrodynamics, p. 139. London: Longmans 1964.
43. Stephen, M. J.: J. Chem. Phys. 40, 669 (1964).
44. Lax, M.: Phys. Rev. 172, 350 (1968).
45. Haken, H., Weidlich, W.: Z. Physik 205, 96 (1967).
46. Lax, M.: In: Chretien et al. (Eds.): Statistical physics, phase transitions and superfluidity, Vol. 2, p. 269. New York: Gordon and Breach 1968.
47. Bethe, H. A.: Phys. Rev. 72, 339 (1947).
48. Bullough, R., Caudery, P. J.: J. Phys. A 4, L 41 (1971).
49. Rehler, N., Eberly, J. H.: Phys. Rev. A 3, 1735 (1971).
50. Stratonovich, R. L.: Topics in the theory of random noise, Vol. I. New York: Gordon and Breach 1963.
51. Lax, M.: Rev. Mod. Phys. 38, 541 (1966).
52. Lax, M.: Phys. Rev. 145, 110 (1966).
53. Senitzky, I. R.: Phys. Rev. 161, 165 (1967).
54. Van der Pol, B.: Phil. Mag. 3, 65 (1927).
55. Haken, H.: In: Flügge, S. (Ed.): Laser theory, Vol. XXV/2C. New York: Springer 1970 Handbuch der Physik.
56. Sudarshan, E. C. G.: Phys. Rev. Letters 10, 277 (1963). — Glauber, R. J.: Phys. Rev. 131, 2766 (1963).
57. Senitzky, I. R.: Phys. Rev. 131, 2827 (1963), and references therein.
58. Senitzky, I. R.: Phys. Rev. A 6, 1175 (1972).
59. Haken, H.: Z. Physik 181, 96 (1964); 182, 346 (1965); 190, 327 (1966).
60. Bullough, R.: In: Mandel, L., Wolf, E. (Eds.): Proc. Third Rochester Conference on Coherence and Quantum Optics, p. 121. New York: Plenum Publishing Corporation 1973.
61. Bullough, R., Saunders, R.: to be published.
62. Series, G. W.: In: Skalinski, T. (Ed.): Optical pumping and atomic line shape, p. 25. Warsaw: Panstwowe Wdawnictow Naukowe 1969.
63. Born, M., Wolf, E.: Principles of optics, p. 81. New York: Pergamon Press 1970.
64. Willis, C. R.: J. Math. Phys. 5, 1241 (1964); 6, 1984 (1965).
65. Mori, H.: Prog. Theoret. Phys. (Japan) 33, 423 (1965).
66. Agarwal, G. S.: Nuovo Cimento Letters 2, 49 (1971).
67. Slater, J. C.: Quantum theory of atomic structure, Vol. 2, Appendix 31. New York: McGraw-Hill 1960.
68. Friedberg, R., Hartmann, S. R., Manassah, J. T.: Phys. Reports 7 C, 101 (1973).
69. Friedberg, R., Hartmann, S. R.: Opt. Commun. 2, 301 (1970).
70. Ernst, V., Stehle, P.: Phys. Rev. 176, 1456 (1968).
71. Bullough, R., Saunders, R.: J. Phys. A 6, 1348, 1360 (1973).
72. Lehmberg, R.: Phys. Rev. A 2, 889 (1970).
73. Varfolomeev, A. A.: Sov. Phys. JETP 32, 926 (1971).
74. Lama, W. L., Jodoin, R., Mandel, L.: Am. J. Phys. 40, 32 (1972).
75. Zakowicz, W.: Phys. Letters 32 A, 87 (1970). — Katriel, J., Adam, G.: Phys. Letters 33 A, 190 (1970).

76. Arecchi, F.T., Kim, D.: Opt. Commun. **2**, 324 (1970).
77. Bonifacio, R., Schwendimann, P., Haake, F.: Phys. Rev. A **4**, 302 (1971).
78. Bonifacio, R., Schwendimann, P., Haake, F.: Phys. Rev. A **4**, 854 (1971).
79. Bonifacio, R., Gronchi, M.: Nuovo Cimento Letters **1**, 1105 (1971).
80. Haake, F., Glauber, R.J.: Phys. Rev. A **5**, 1457 (1972).
81. Ponte Goncalves, A.M., Tallet, A.: Phys. Rev. A **4**, 1319 (1971). — Degiorgio, V., Ghielmetti, F.: Phys. Rev. A **4**, 2415 (1971).
82. Bailey, N.T.J.: The elements of stochastic processes with applications to the natural sciences, New York: Wiley and Sons.
83. Dillard, M., Robl, H.R.: Phys. Rev. **184**, 312 (1969).
84. Dialetis, D.: Phys. Rev. A **2**, 599 (1970).
85. Friedberg, R., Hartmann, S.R., Manassah, J.T.: Phys. Letters **40** A, 365 (1972).
86. Stroud, C.R., Eberly, J.H., Lama, W.L., Mandel, L.: Phys. Rev. A **5**, 1094 (1972).
87. DeGiorgio, V.: Oopt. Commun. **2**, 362 (1971).
88. Gordon, J.P., Aslaksen, E.W.: IEEE J. Quant. Electr. QE **7**, 428 (1970).
89. Abramowitz, M., Stegun, I.A.: Handbook of Mathematical Function, p. 228. New York: Dover 1964.
90. Louisell, W.H.: In: Glauber, R.J. (Ed.): Quantum optics, p. 680. New York: Academic Press 1969.
91. Kocher, C.A.: Ann. Phys. **65**, 1 (1971).
92. Cho, Y.C.: Thesis, Massachusetts Institute of Technology, to be published.
93. Nesbet, R.K.: Phys. Rev. Letters **27**, 553 (1971).
94. Walsch, J.E.: Phys. Rev. Letters **27**, 208 (1971).
95. Barton, G.: Phys. Rev. A **5**, 468 (1972).
96. Lehmberg, R.H.: Opt. Commun. **5**, 152 (1972).—Agarwal, G.S.: Ref. [27], § XI.
97. Mollow, B.R., Miller, M.M.: Ann. Phys. **52**, 464 (1969). — Mollow, B.R.: Phys. Rev. **188**, 1969 (1969); A **5**, 1522 (1972). — See also Newstein, M.C.: Phys. Rev. **167**, 89 (1968). — Keller, O.A., Robiscoe, R.T.: Phys. Rev. **188**, 82 (1969).
98. Gush, R., Gush, H.P.: Phys. Rev. A **6**, 129 (1972).
99. Stroud, C.R.: Phys. Rev. A **3**, 1044 (1971).
100. Fain, V.M., Khanin, Ya.I.: Quantum electronics, Vol. I. Cambridge: MIT Press, 1969.
101. Bloch, F., Siegert, A.: Phys. Rev. **57**, 522 (1940).
102. Autler, S.H., Townes, C.H.: Phys. Rev. **100**, 703 (1955).
103. Cohen-Tannoudji, C.: In: Lévy, M. (Ed.): Cargèse lectures in physics, Vol. 2, p. 347. New York: Gordon & Breach 1968.
104. Stenholm S.: J. Phys. B **5**, 878 (1972).
105. Bogoliubov, N.N., Mitropolsky, J.A.: Asymptotic methods in the theory of non-linear oscillations, Chap. 5. Delhi: Hindustan Publishing Co. 1961.
106. Walls, D.F.: Phys. Letters **42** A, 217 (1972).
107. Arnous, E., Heitler, W.: Proc. Roy. Soc. (Lond.) A **220**, 229 (1953).
108. Knight, P.L., Allen, L.: Phys. Rev. A **7**, 368 (1973).
109. Casimir, H.B.G., Polder, P.: Phys. Rev. **73**, 360 (1948).
110. Agarwal, G.S.: Z. Physik **258**, 409 (1973).
111. Haken, H.: Z. Physik **263**, 267 (1973).

Dr. G. S. Agarwal
Tata Institute of Fundamental Research
Homi Bhabha Road, Colaba
Bombay-5, INDIA

Added in Proofs

We would like to mention two interesting developments which have taken place since this article was submitted for publication.

(i) R. Bonifacio, G. P. Banfi and P. Schwendimann have made considerable progress towards the solution of (6.53) for the case of large samples.

(ii) The present author has developed a theory of coherence effects which arise in spontaneous emission due to the presence of dielectric and conducting interfaces. This theory proceeds along lines similar to that of Chapters 6–10.

SPRINGER TRACTS IN MODERN PHYSICS

Ergebnisse der exakten Naturwissenschaften

Quantum Statistics

Graham, R.: Statistical Theory of Instabilities in Stationary Nonequilibrium Systems with Applications to Lasers and Nonlinear Optics (Vol. 66)

Haake, F.: Statistical Treatment of Open Systems by Generalized Master Equations (Vol. 66)

Agarwal, G. S.: Quantum Statistical Theories of Spontaneous Emission and their Relation to Other Approaches (Vol. 70)

Semiconductors

Feitknecht, J.: Silicon Carbide as a Semiconductor (Vol. 58)

Grosse, P.: Die Festkörpereigenschaften von Tellur (Vol. 48)

Schnakenberg, J.: Electron-Phonon Interaction and Boltzmann Equation in Narrow Band Semiconductors (Vol. 51)

Superconductivity

Lüders, G., Usadel, K.-D.: The Method of the Correlation Function in Superconductivity Theory (Vol. 56)

X-Ray, Neutron-, Electron-Scattering

Steeb, S.: Evaluation of Atomic Distribution in Liquid Metals and Alloys by Means of X-Ray, Neutron and Electron Diffraction (Vol. 47)

Springer, T.: Quasi-Elastic Scattering of Neutrons for the Investigation of Diffusive Motions in Solids and Liquids (Vol. 64)

To Appear in Volume 71

Überall, H.: Study of Nuclear Structure by Muon Capture

Singer, P.: Emission of Particles Following Muon Capture in Intermediate and Heavy Nuclei

Levinger, J. S.: The Two and Three Body Problem

To Appear in Forthcoming Volumes:

Langbein, D.: Theory of van der Waals Attraction

Brandmüller, J., Claus, R.: Light Scattering on Optical Phonons and Polaritons

Bauer, G.: Determination of Electron Temperatures and of Hot-Electron Distribution Functions in Semiconductors